Beginner's Guide to

Reading Schematics

2nd Edition

Beginner's Guide to Reading Schematics

2nd Edition

Robert J. Traister
Anna L. Lisk

TAB Books
Division of McGraw-Hill
New York San Francisco Washington, D.C. Auckland Bogotá
Caracas Lisbon London Madrid Mexico City Milan
Montreal New Delhi San Juan Singapore
Sydney Tokyo Toronto

pbk 23 24 25 26 27 28 29 30 FGR/FGR 0 9
hc 3 4 5 6 7 8 9 10 FGR/FGR 9 9 9 8 6 5 4 3 2

Library of Congress Cataloging-in-Publication Data

Traister, Robert J.
Beginner's guide to reading schematics / by Robert J. Traister and Anna L. Lisk.
p. cm.
Includes index.
ISBN 0-8306-7632-5 (p) ISBN 0-8306-8632-0
1. Electronics—Charts, diagrams, etc. I. Lisk, Anna L.
II. Title.
TK7866.T7 1991
621.381—dc20 91-265
CIP

Acquisitions Editor: Roland S. Phelps
Technical Editor: Andrew Yoder
Book Design: Jaclyn J. Boone
Director of Production: Katherine G. Brown
Cover photograph: Susan Riley, Harrisonburg, Va.

Contents

Introduction

MANY PEOPLE SHY AWAY FROM ELECTRONIC PURSUITS because they think reading and drawing schematic diagrams will be complex and difficult. The fear of the unknown, however, is quickly erased when just a bit of knowledge is conveyed about this subject area. Refusing to enter an electronics pursuit because of schematic diagrams is equivalent to refusing to go swimming because of a fear of lifeguards The lifeguard is put there to make swimming safer and easier. This is the prime purpose of schematic diagrams as well.

A schematic diagram is a road map of an electronic circuit; with it you have your own personal guide to understanding simple circuits, complex circuits, and even massive systems. Reading a schematic diagram is no more difficult than reading a road map once you have the proper background.

The purpose of this book is to provide the basic information you need to begin exploring electronic circuits and then to show you how to use that knowledge for circuit analysis, troubleshooting, and repair. This book explains in clear language the reason for schematic diagrams, how each symbol is derived and used, and how individual symbols are combined to make electronic circuits.

Before you can even dabble in electronic circuits, it is mandatory that you have a working knowledge of schematic diagrams and their uses. Learning to read schematic diagrams and to use them to analyze electronic circuits is one of the simplest parts of electronic experimentation, design, and troubleshooting. Schematic diagrams are "human-engineered" to allow the interfacing of human deductive powers with electronic circuits.

This book will take you through the basics and provide enough information to allow you to continue to perfect your ability to delve deeply into electronic circuits with an understanding of function and design. This is the first step on your journey to electronics proficiency.

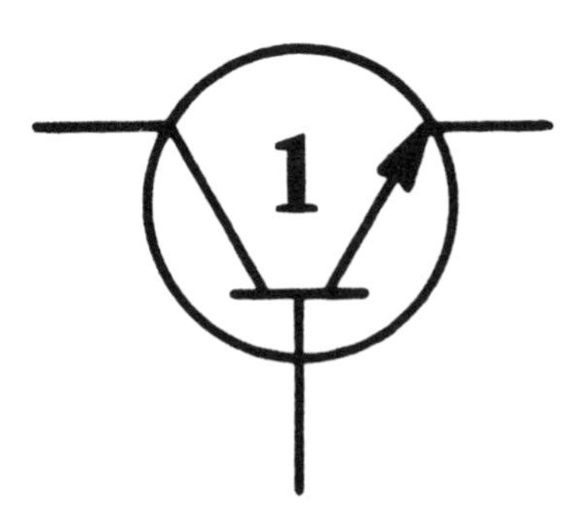

Types of electronic diagrams

THERE ARE THREE BASIC TYPES OF ELECTRONIC DIAGRAMS: block, schematic, and pictorial. Each of these types is discussed separately in this book. In using electronic diagrams to completely understand a circuit or device, the most typical progression essentially goes in that order. A *block diagram* gives a basic overview of how the main circuits within a device interact. Each main function is represented with a block (rectangle or other shape, depending on the application), and the interconnecting lines or arrows show the relationships between each.

The second main type is the *schematic diagram*. Schematic diagrams contain every component that makes up a circuit via various symbols. These symbols and how they interconnect is a large part of the focus of this book.

Pictorial diagrams, sometimes called *layout diagrams*, show the physical relationships of how the components are arranged to facilitate finding components to test or replace them Therefore, when troubleshooting an unfamiliar electronic circuit, you usually start with the block diagram to find where the trouble might be coming from. Then refer to that schematic or that part of the schematic to find the faulty component(s). Finally, the pictorial provides the information of where the faulty component(s) can be found to be tested and/or replaced.

Block diagrams

Block diagrams are used in conjunction with schematics to aid circuit comprehension and to accelerate troubleshooting procedures. Each block is assumed to represent all schematic symbols related to that part of the circuit and represents it as a block. Each block is labeled with a description of the circuit it represents. The block diagram does little or nothing to explain the actual makeup of the circuit it represents. Instead they are functional in nature; they describe the circuit function rather than depicting actual components. Once you have a basic understanding of the principal circuit functions via the block diagram, you can then consult the schematic for more practical details to troubleshoot or construct a circuit.

To understand how block diagrams might be used, consider the following examples. Suppose you would like to design a circuit. You can simplify matters by beginning with a block diagram that would show in block form all of the circuits needed to complete the project. From that, you can transform each block into a schematic diagram Eventually, you will have devised a complete schematic that replaces all of the blocks.

A second method works the other way around. For example, say you have a complicated schematic that you are trying to troubleshoot. With every component shown, it can be difficult to determine which part contains the problem area. A block diagram would provide a clearer understanding of how each part of the circuit operates in conjunction with the others. Once the possibly troublesome area is located, you can then return to the schematic for more details. Chapter 2 covers block diagrams thoroughly.

Schematic drawings

A schematic diagram is a map of an electronic circuit showing every component and how they are interconnected. According to *Webster's*, *schematic* means "of or relating to a scheme; diagrammatic." Therefore, any diagram that depicts a scheme—be it electronic, electrical, physiological, or whatever—can be classified as a schematic drawing. Most people depend upon schematic drawings every day to give the whole picture of a scheme or event within a confined medium. This aids the understanding of the scheme when viewed in its entirety.

One of the most common schematic diagrams is used by nearly everyone who has ever driven an automobile. A ***road map*** is a diagrammatic form of representing an entire scheme. The scheme might involve the paths of travel within a small locality, within a state, or even within several states. Like a schematic diagram of an electronic circuit, the road map shows all the components that are relative to the particular travel scheme it addresses. Motorists will make up their own schemes, which are often small portions of the total scheme included in the road map. Likewise, electronic schematic diagrams can be used to show the whole scheme and also to allow the technician to extrapolate the section, which fits the scheme in mind.

Using the road map as a comparison to electronic schematic drawings, let's assume you wish to travel in your automobile from point A to point B. We can say that the road map will list all of the towns and cities that lie between these two points. By comparison, it can be said that a schematic diagram will list all components between a similar point A and point B in the circuit. However, both schematic diagrams indicate much more. It's not enough to know which towns or cities lie between these two points to get an idea of the overall scheme of things. Indeed, we could easily write down the names of these locations, in which case we would not have to resort to a diagram at all. From the electronics standpoint, we could do the same thing by simply providing a list of the components that were used to build a certain circuit such as:

- 120-ohm resistor
- 1000-ohm resistor

- pnp transistor
- .47 μF capacitor
- 2 feet of hookup wire
- 1.5-volt battery
- switch

Now, what has this list told us about the circuit? Not much, really. We do know the components involved in building it, but unfortunately, we don't know what it is that was built.

Any schematic drawing must not only indicate all components necessary to make a specific scheme but also *how* these components are interrelated—how they are connected. The road map interconnects the various towns, cities, and other trip components by lines that represent streets and highways. A line that indicates a secondary road is different from one that is used to represent a four-lane highway. With a bit of practice, we can tell which lines indicate which roads. Likewise, an electronic schematic drawing uses lines to indicate a standard conductor; other types will be used to represent a cable. In both cases, when the interconnecting lines are added, a relationship is established between the connected components. Then too, the physical relationship of one component of a road map to others tells us something as well. This relationship is not as true in electronic diagrams, but they do provide a type of relationship that is enforced by interconnecting lines representing conductors.

Symbols

Now it becomes necessary to describe *how* a schematic diagram displays the scheme of a system. It displays through symbology. The lines that indicate roadways are symbols. Symbols are used as opposed to pictorial drawings. A single black line that might indicate Route 522 in no way resembles the actual appearance of this highway. It is enough for us to know that a black line symbolizes Route 522. We can make up the rest in our own minds. If it was necessary to provide a pictorial drawing of this route, road maps would be a hundred times larger and probably ten thousand times more difficult to read.

The same is true of towns, cities, railroad tracks, airports, and a thousand other features found on a standard road map. You cannot pictorially represent them in a practical manner. Instead, these components are represented by symbols. A key to the symbols used is often provided on the map. It shows the symbol and explains in plain language what each means. If a small airplane drawn on the map indicates an airport and this fact is known to us, then each time we encounter the airplane symbol, we will know that an airport exists at this site. Again, we are able to visualize the airport in our minds using the tiny drawing as a stimulus. Symbology involves the depiction of a physical object (in this case only) by means of another physical object (the miniature airplane) for the purpose of practical representation. In individual cases, the makeup of the symbol is not that important.

However, a road map must contain many different symbols. Each is usually human engineered to be logical to the human mind. For instance, when you see a miniature airplane on a road map, you will be inclined to think that this area had something to do with airplanes, so a detailed explanation would not be necessary. If, on the other hand, you used the ridiculous example of a beer bottle to represent the same thing, anyone who did not read the key would certainly not be inclined to think of an airport. Since many different symbols would be used, it is mandatory that whenever possible, each would be presented in a logical manner.

Logic will only take it so far, however. This phenomenon is especially true in electronics diagramming, where the actual appearance of a component might vary greatly from manufacturer to manufacturer. A certain logic does apply, however, after a basic groundwork has been laid in electronics symbology. For example, a circle is used to indicate a vacuum tube. Other symbols are used inside this circle to represent the many tube electrodes. A tube is an active device, capable of producing an output that is of higher amplitude than the signal at its input The same can be said of a transistor, which was developed many years after vacuum tubes. Since a circle with electrode symbols had been used for many years to represent vacuum tubes and as a result transistors were developed as active devices to take the place of some tubes, the schematic symbol for the transistor also started with a circle. Electrode symbols were inserted into this circle as before, but the symbols here were different from tube elements, so the two types of devices could be easily distinguished. The logic here is based around the circle symbol. Transistors accomplish many of the same functions in electronic circuits as vacuum tubes do, so symbolically they are quite similar as to circuit function. It would not be logical to develop a symbol that was far removed from the vacuum tube symbol, which had been used for decades.

There are certain inconsistencies, however. Circles sometimes make up a part of an electrical symbol, indicating solid-state devices that are *not* symbolically equivalent to tubes and transistors. A zener diode, for example, is often indicated by a circle with a special diode symbol at its center. A zener diode is not a transistor, and the electrode symbol at the center clearly indicates that this is not a transistor. Fortunately, the circle symbols today are most often used to indicate either a vacuum tube or some sort of solid-state device. This is not always the case, but it is often so. Chapter 3 will describe the various components which are represented schematically and will explain how each symbol is to appear and how it is drawn.

Interconnections

To further explain how schematic diagrams are used, we can take a single component, a pnp transistor. This device has three electrode elements, and although there are many thousands of different varieties of pnp transistors, they all will be drawn symbolically in the same manner. You must remember that there are thousands of different circuits into which a single pnp transistor can be inserted. Schematics are really needed to indicate *how* the transistor is connected into the circuit, what other components are used in conjunction with this device, and what other circuit portions depend upon this device for overall operation. A transistor, for example, can be used as a solid-state switch, an amplifier, an impedance-matching device, etc. One type of transistor can serve all of these purposes. Therefore, if a transistor is used in one circuit as an amplifier, you cannot say that this transistor is used as an amplifier only. As a matter of fact, you could pull this transistor out of the amplifier circuit and put it into another one to form a solid-state switch. By knowing the type of component alone, you cannot know how it is used in a circuit until you can get an overall view of all connections and interconnections. This can't be done, in most instances, by examining the physical circuit. You need a road map, a schematic diagram, to show you all the connections that were made to form the circuit.

A schematic is needed because the human brain cannot retain all of the input data that is fed to it by the eyes when scanning a small portion of a physical circuit. As a practical nonelectronic example, let's assume that you are to drive from Washington, DC to Los Angeles, CA. Even if you had made the trip several times before, there's a good chance that you could not remember all of the routes to take and all of the towns and cities that you passed on the way. A road map, however, would give you an overall picture of the entire trip. Since all of the trip data has been collected and presented in a form that can be scanned at a glance, the road map becomes instrumental in allowing you to see the entire trip rather than a piece of it at a time. The schematic diagram does the same thing for a trip through an electronic circuit.

Using the road map and the coast-to-coast trip again as an example, let's assume that you have memorized the entire route. Assume also that one of your prime routes of travel along the way is under construction, and it becomes necessary to take an alternate route. Without a road map, you would not know a separate artery to take, one which would keep you on course as much as possible and eventually return you to the original travel path.

In an electronic circuit, there are many electrical highways and byways. Occasionally, some of these break down, making it necessary to seek out the problem and correct it. Even if you can visualize the circuit in your head as it appears in physical existence, it is nearly impossible to have any idea of the many different routes that are used, one or more of which could be defective. When I speak here of visualizing the circuit, I am not speaking of the schematic equivalent of the circuit, but the actual hard wiring itself. A schematic diagram is necessary to give you an overall picture of the circuit and to show how the various routes and components depend upon other routes and components. When you can see how the overall circuit depends upon each individual circuit leg and component, it then becomes easier to diagnose the problem and to effect repairs.

Symbology

It is often difficult to fully explain schematic diagramming to individuals who are just starting out in electronics. One must think of this form of symbology as a language. What is a language? It is a system of symbols that are used to communicate ideas. The English language is a symbology with which most of us are quite familiar.

Every word spoken in English or any other verbal language is a complex symbol made from simple symbols called *number* and *letters*. Let's take the word "stop," for example. Without a reference key, it means nothing. However, through learning the symbology from shortly after birth, this word begins to mean something because the infant, who is learning to speak and understand, can compare "stop" to other words, especially to actions. We can even say the word "stop" is symbology within symbology. The communication's intent of this word can also be expressed by the phrase "Do not proceed further." This phrase, however, is still symbology, expressing a mental image of a desired action.

If we could all communicate by telepathy, then symbology would not be necessary. Some would argue with this, saying that we all think in our own language, language being symbology. I do not believe this to be true. Thinking is done on a far speedier level and is identical from human to human, regardless of what language or languages he understands. A newborn baby, for instance, speaks and understands no language whatsoever. However, whether that baby was born in the United States, South Africa, Asia, or wherever, thought processes do take place.

The baby knows when it is hungry, in pain, frightened, etc. It needs no language to comprehend this. It does become necessary to communicate right from the start. For this reason, all newborns communicate in the same language (crying, mostly). As newborns are able to comprehend more and more of their environment through improved sensory equipment (eyes, ears, etc.) more data is collected.

The environment then plays a role in allowing the baby to comprehend communications symbology. Here, the various languages come into play, with different societies using different verbal symbols to express simple and complex mental processes. The human brain still carries on the same nonlinguistic thought processes as before because thinking in terms of symbols would take far too much time and memory storage area.

The brain does, however, allow the human being to transpose complex thoughts into a language. When a child is about to step in front of a speeding automobile, if the brain had to handle the words "automobile, speed, death, child" and literally millions of other criteria symbolically, we human beings would spend all of our lives waiting for it to deliver the correct processed information. Rather, the brain scans all that is received by the sensory organs in real time and then sums it up into a single symbol which can be used for communcation. The symbol is the word "stop!" When that word is communicated to the child, a similar process takes place in his or her brain, having been triggered by the uttered symbol.

All languages do not involve the spoken word, however. We've all heard of Indian sign language, whereby the arms and hands are used to communicate ideas. In most instances, an entire communicating language of visual symbols is not as efficient (to human beings) as one composed of words and visual symbols. This is partially as a result of the way human beings speak words. Using the symbol "stop" again, we know that this can be uttered in many different ways The word in itself means something, but the way it is spoken augments the meaning. The tone of voice, inflection and certainly the volume at which it is uttered car modify the basic meaning. It is sometimes difficult to do the equivalent by visual means only. We have arrived at some universal methods of modifying visual symbols. To many of us, the color red denotes danger or at least something that should be given immediate attention. Often, however, this color is used in conjunction with the visual symbol for a spoken word.

As far as the symbology of schematic drawing is concerned, this visual language was caused by the makeup of human beings. It does not lend itself readily to any form of oral symbology. Our senses along with our central processor, the brain, render us less than proficient at mentally conceiving all of the workings of electronic circuits by dealing with them directly. Therefore, it is necessary to accept data a small step at a time, compiling it in hard-copy form (through symbology) and providing a hard-copy readout—the schematic diagram. In order for us to understand what takes place in the entire circuit, all of this collected information must be presented at one time and in a form whereby we can see of the steps that went into its making at one time. This is how schematic diagrams are prepared. This method can be likened to the "connect-the-dots" drawings that are offered in many children's workbooks. Individually, the dots mean nothing, but once they are arranged in logical form and connected by lines, we get an overall picture. One cannot, however, deny that the picture was formed starting with a few dots. The dots themselves are unimportant, but their relationships to each other and to the order in which they are connected means everything.

Don't let this discussion on symbology lead you to think that schematic diagramming is a highly complex field. It is not; schematic drawings are provided to make things simple.

We use symbology every day and take it for granted. Therefore, schematic diagrams are something we can logically take to, and in a short period of time.

The chapters in this book start with the basics of schematic diagramming, the symbols, and take you through to simple circuits, and finally to complex ones. Schematic symbols and diagrams are designed for humans; therefore, human logic is a prime factor in determining which symbols mean what. Anyone who is able to read and do simple math can learn to read and draw schematic diagrams.

Pictorial diagrams

Recognizing the difference between schematic and pictorial diagrams is an important step. Schematic diagrams are symbolic representations of electronic circuits and pictorials are physical representations. Pictorial drawings show the actual proportional sizes of the components, while schematic diagrams depict the circuit with the components shown by symbol, without regard to size or shape.

Once you understand block diagrams and the details of schematics, this book concludes with an explanation and demonstrations of how pictorial diagrams are an important final step in analyzing, understanding, troubleshooting, and repairing electronic circuitry.

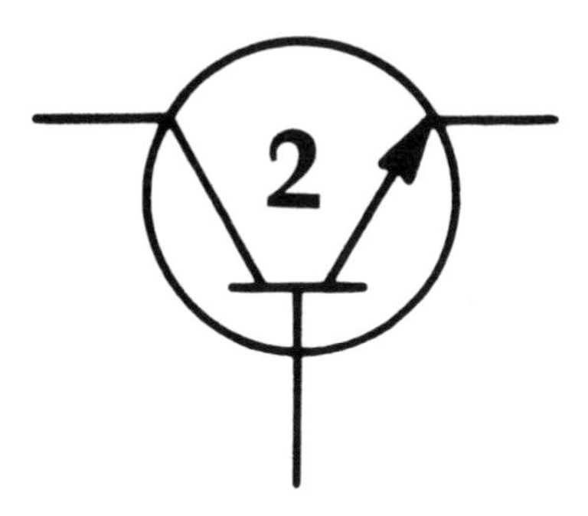

Block diagrams

AS EXPLAINED IN CHAPTER 1, BLOCK DIAGRAMS HELP SHOW how an electronic circuit is constructed. They are used to show a simplified version of a circuit by separating the main parts and showing how they are interconnected.

Figure 2-1 shows a simple block diagram of an ac-to-dc converter. At left are the input terminals, which accept an ac signal. In sequence, the signal passes through the transformer, the rectifier, and the filter before arriving at the output as a dc signal. In this case, the lines that connect each block do not have arrows because the progression is assumed to go from left to right. The main reason for this assumption is because the input is at the left and the output is on the right. In more complicated diagrams, arrows are required to show which block is affecting which. This circuit is discussed later in this book in both schematic and pictorial form.

Function

Block diagrams can be handled in many different manners. Sometimes they are used to indicate interconnections between various pieces of equipment. When drawn as shown previously in this discussion, they are often called *functional diagrams* because they indicate the basic functioning of the

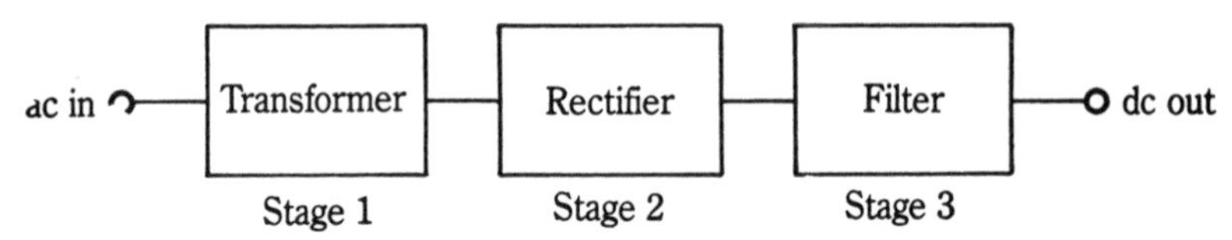

2-1 Simple block diagram of an ac-to-dc converter. The signal path flows from left to right.

electronic circuit. The functional diagram allows for an easy explanation as to how the device operates and will lead into further explanations provided by a schematic diagram of the complete system.

There are several ways block diagrams can be used. First of all, someone who is trying to arrive at a schematic diagram for a complex electronic circuit that must be designed more or less from scratch might decide to start with a block diagram. This diagram would show, in block form, all of the circuit sections needed to arrive at a functioning device. Then the designer would seek out schematic diagrams of circuits that could fill each block. In most instances, these circuits will have to be modified to blend with the overall system. The first block in the diagram would then be substituted by the schematic diagram of the circuit it specifies. The schematic designer would move through the blocks according to functional order, designing schematic diagrams that can be used to build functioning circuit sections. As soon as the final block is filled in schematically, the device is complete. Eventually, a total theoretical circuit has been designed on paper and can be put in a finished schematic form, removing all blocks.

Another way of using block diagrams starts with possessing a finished schematic diagram. Assume that the schematic is very complex and that the equipment whose circuit it represents is malfunctioning. Although schematic diagrams do describe the functioning of an electronic circuit, they are not as clear and basic as a functional block diagram. The technician would then laboriously identify each circuit section and draw it in block form. When finished, this drawing would provide a clearer understanding of how each circuit section operates in conjunction with all others. Using this method, one or more circuit sections can be identified as a possible trouble area. At this point, the original schematic drawing would be referenced again and tests would be made in the indicated areas.

In practice, you will often encounter block diagrams. If presented without accompanying schematic drawings, these diagrams will be used mainly to describe the basic functional operation of a class of circuit or device. The block diagram is used where a literal interpretation of individual circuit functions is not necessary. For example, we can describe the operation of a specific type of radio transmitter (amplitude modulated, continuous wave, frequency modulated, etc.) by means of a block diagram. This diagram would be applicable to nearly all radio transmitters of certain basic design. Now, no two types of radio transmitters built by different manufacturers are exactly alike, but all of them would contain the same basic circuit sections as far as function is concerned. One type of oscillator might work differently from another type, but they all perform the same function. When individual differences must be represented, then the actual schematic drawing is used.

Observe the block diagram in Fig. 2-2. This diagram illustrates the various parts of a strobe light circuit. The signal path is from left to right. Let's go through the diagram block by block to understand how it works. Later in this book, the circuit is depicted schematically and pictorially.

The input signal enters at the left of the diagram, which is 117 Vac. The signal "splits" and goes to both a fuse and a

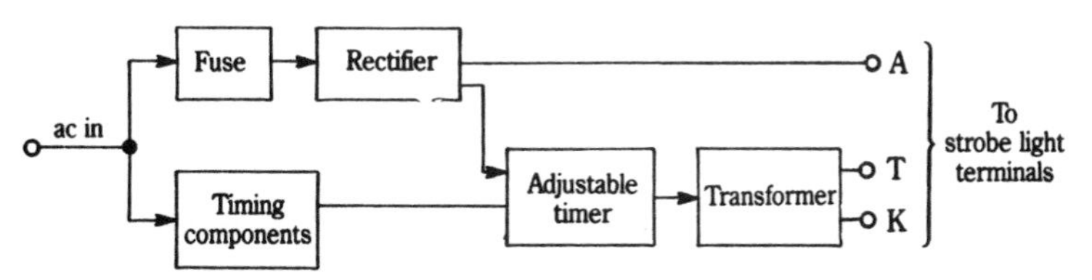

2-2 Simple block diagram of a circuit that powers a strobe light. Arrows are shown to illustrate the direction of the signal path in more complicated circuits.

combination of components that provide a timing combination. The top path, where the fuse is, then follows to a diode rectifier, and the rectifier's output passes directly to one terminal of the three-terminal strobe lamp. The rectifier also outputs to an adjuster that provides a variable rate of blink for the light. The output from that adjuster goes to a transformer, which then provides the remaining two outputs required to power the strobe light.

Another example is the block diagram in Fig. 2-3. This circuit is slightly more complicated—a power supply that produces several different voltage outputs to power the various requirements of a VCR. Walking through this circuit from left (the input) to right (most of the outputs), note that the circuit is powered with 120 Vac (standard house current). The voltage then goes through a filter and splits. Part of the signal goes to the "lower" transformer where a 16-Vac source and a 3-Vac source are output. An output that provides circuit ground also comes out of this transformer.

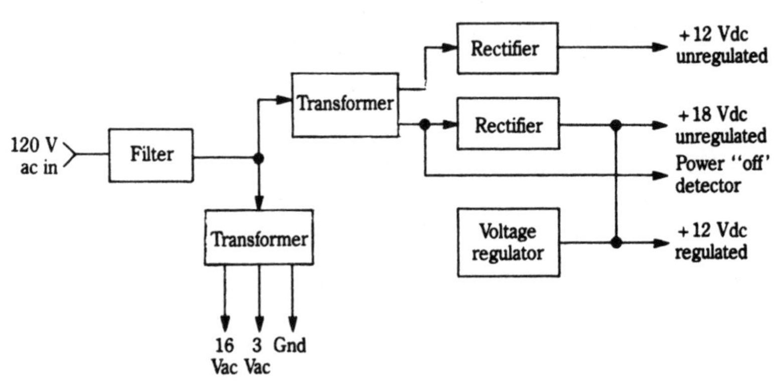

2-3 Block diagram for a power supply that produces seven different outputs.

From the filter, the input voltage goes to another transformer that derives the dc voltages. One output of the transformer goes to a rectifier that provides 12 Vdc unregulated. The other transformer output goes to a separate rectifier that provides 18 Vdc unregulated. This transformer output also serves as a diagnostic detector for a "power-off" condition.

That line is further tapped to join with the output of the voltage regulator to give 12 Vdc regulated. This circuit is also discussed later in this book in both schematic and pictorial form.

Block diagrams are extremely simple to draw and consist of squares or rectangles and sometimes triangles (used to represent a circuit block that is built around an IC amplifier). The block diagram of an AM transmitter shows the interconnections between each of the simple circuits (Fig. 2-4).

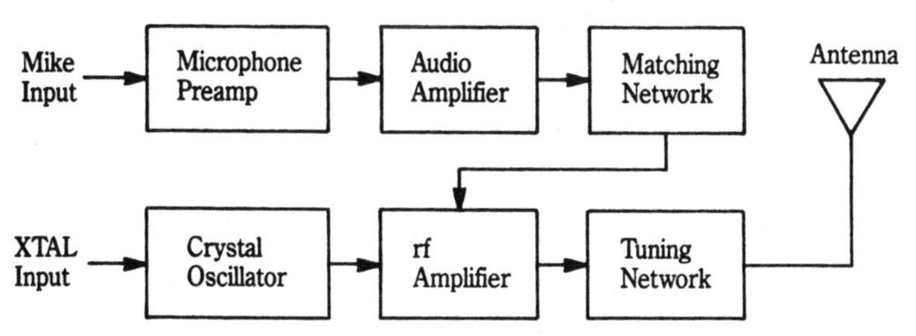

2-4 Block diagram of a typical AM transmitter.

The microphone preamplifier stage is connected to the input of the audio amplifier stage (note the direction of the arrow). The output of the audio amplifier is connected to the matching network, which in turn is connected to the rf amplifier section. The crystal oscillator is also connected to the rf amplifier section, whose output leads into the rf tuning network. Only one connection is between the audio section of the circuit and the rf section, and that is between the matching network and the rf amplifier. This block diagram, then, shows a basic sequence of events or sequence of paths throughout the entire circuit. The circuit is further explained in Chapter 5.

Flowcharts

Block diagrams are commonly used to describe the functioning of electronic circuits, but in the electronics world of computers, another form of diagramming is needed to display the functioning of a program. This system is called *flowcharting* and is similar to block diagram representation, except the symbology is applied to the different basic sections of the computer program.

The flowchart is a highly useful tool in computer programming and is a graphic representation of the paths that a computer program will take. Flowcharts are often prepared in conjunction with the generation of specifications and are modified as the requirements change to fit within the particular constraints of the overall system.

For complex problems, a formal written specification might be necessary to ensure that everyone involved understands and agrees on what the problem is and what the results of the program should be. To illustrate this, let's assume that a teacher wants a program that will determine a student's grade by calculating an average from grades the student received over a grading period. The teacher will supply the grades to the program as input. Only the average grade is needed as an output. Now, we can make an orderly list of what the program has to do:

- Input the individual grades.
- Add the grade values together to find their sum.
- Divide the sum by the number of grades to find the average grade.
- Print out the average grade.

We can also prepare a flowchart of the program, as shown in Fig. 2-5. As can be seen, a flowchart graphically presents the structure of a complex program so that the relationship between parts can be easily understood. When the flow of control is complicated by many different paths that result from many decisions, a good flowchart can help the programmer sort things out. The flowchart is often useful as a thinking-out tool to understand the problem and to aid in program design. At this stage, the flowchart symbols should have English narrative descriptions rather than programming language statements, since we want to describe what is to be done, not how it is to be accomplished. At a later stage, if formal flowcharts are required for documentation, the flowchart can contain

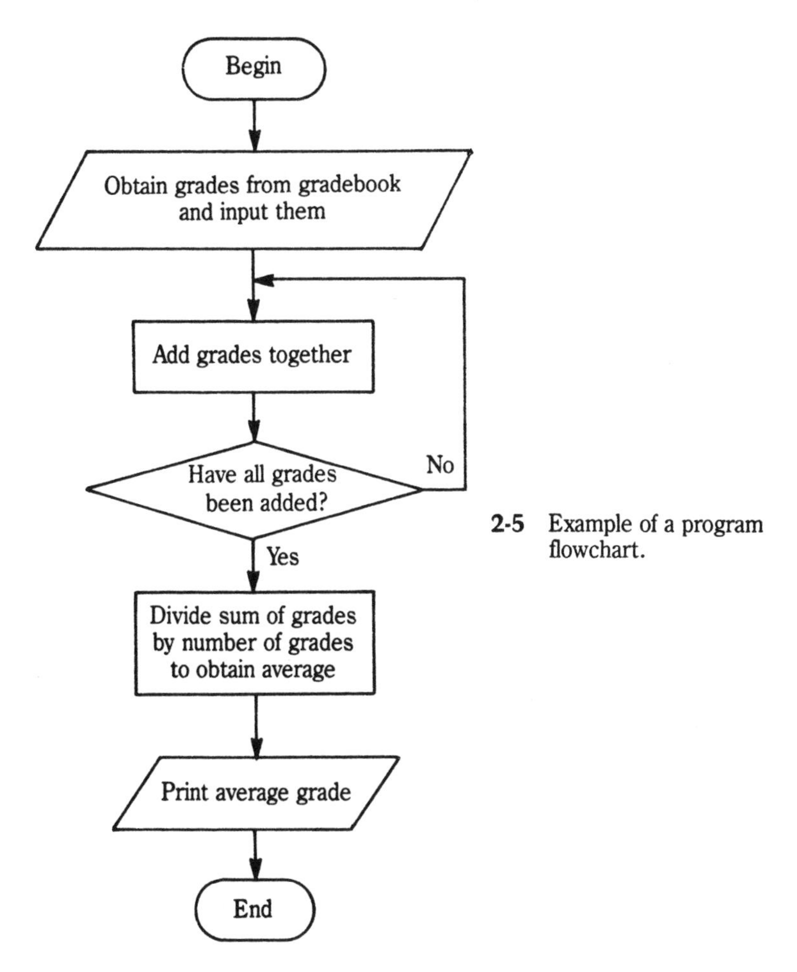

2-5 Example of a program flowchart.

statements in a program language. These flowcharts can be most helpful to another person who at some future time might need to understand the program.

Preparing a formal flowchart is time-consuming, and modifying a flowchart to incorporate changes is often quite difficult. Because of this, some programmers express their dislike for this tool, but most will still use it to provide invaluable assistance in understanding a program.

In order to promote uniformity in flowcharts, standard symbols have been adopted by several organizations. The symbols of the United States of America Standards Institute

(USASI) are widely accepted, and some of their most commonly used symbols are shown and defined in Fig. 2-6. The normal direction of flow in a flowchart is from top to bottom and from left to right. Arrowheads on flow lines are used to indicate flow direction, but they are sometimes left off if the flow is in the normal direction.

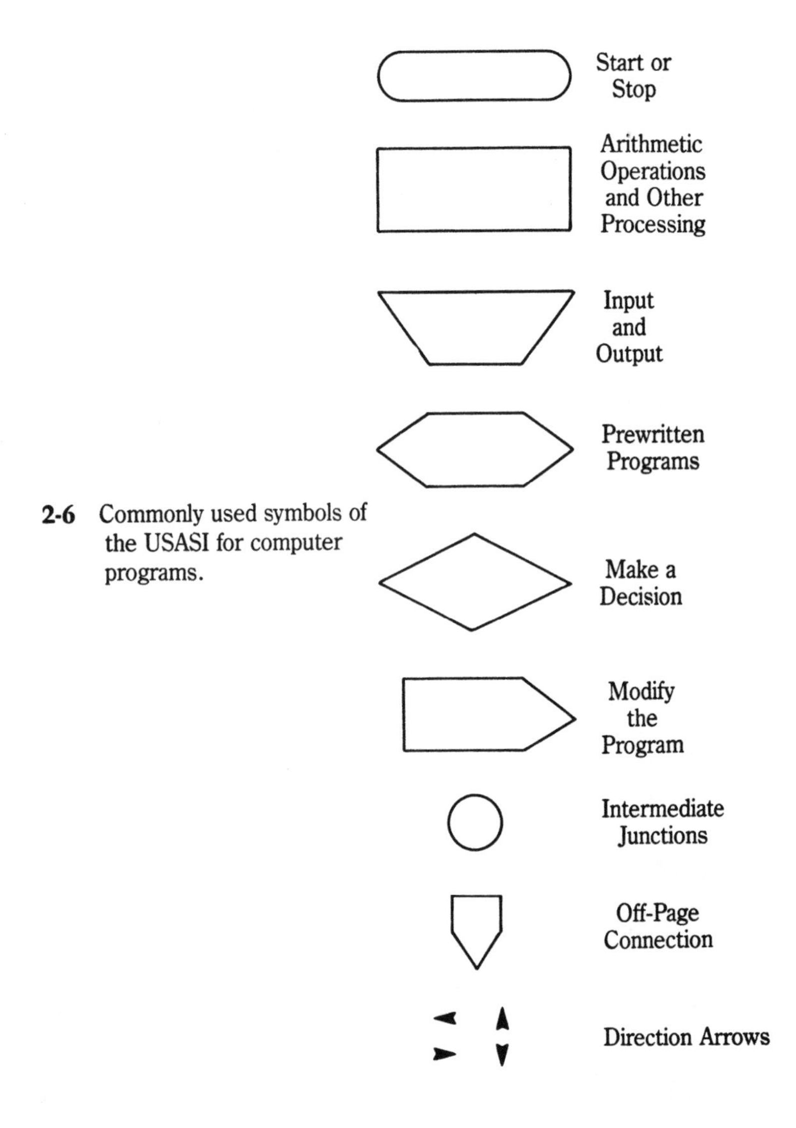

2-6 Commonly used symbols of the USASI for computer programs.

Figure 2-7 shows the flowchart of a program that duplicates punched cards and at the same time prints what is on the card. Let's trace the flow of the program through the flowchart. The program begins at the *start* oval at the top and proceeds in the direction of the arrows at all times. In the first box below *start*, the program reads a card. The next step is to punch the card's contents into a blank card and then print out the contents on the console printer. The program then goes back along the dotted line to the top and reads the next card. The program repeats itself as long as it has cards to read. The part of the program that is done over and over is called a *loop*.

Figure 2-7 used three different symbols to mean different operations. Referring back to Fig. 2-6, which showed some of the more common symbols used in flowcharts, let's look at what these symbols mean. Oval boxes are used to show a start or stop point. Arithmetic operations, such as calculating "height = 1,100 − $16t^2$," are placed in rectangular boxes. Input and output instructions would be placed in an upside-down trapezoid. If we used a program written earlier, we would not bother drawing the flowchart for the inside program. Rather, we would just place the entire program inside

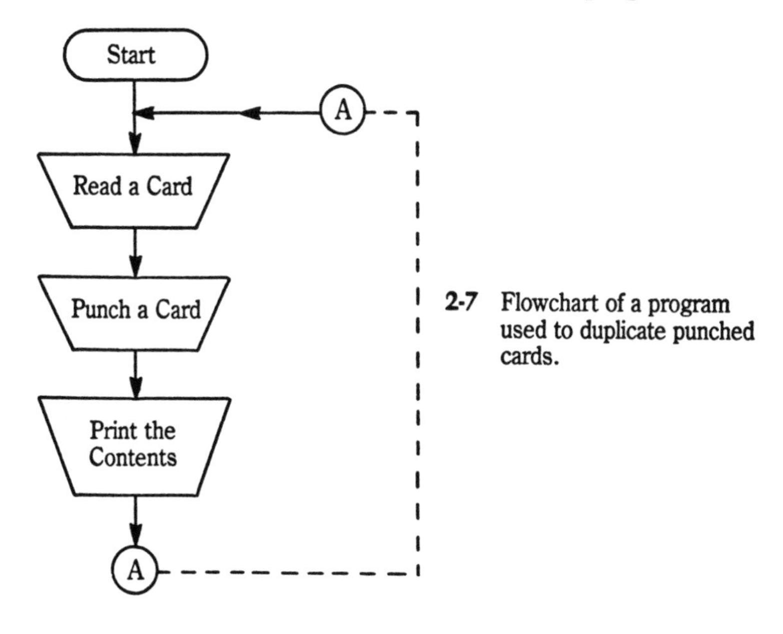

2-7 Flowchart of a program used to duplicate punched cards.

the flattened hexagon. If a box is to be used for making a decision, the diamond shape is used. A five-sided box is used to show a part of the program that changes itself. A small circle identifies a junction point of the program. This point in the program is connected to several places, and we use the intermediate junction symbol to avoid drawing long lines on the flowchart. A small five-sided box is used to show where one page of a flowchart connects to the next (if more than one page is used for the same flowchart). The intermediate junction and off-page connection would further be labeled with a number or letter so that all like symbols with the same letter or number inside are connected together. Finally, arrows show the direction of travel.

Returning to the flowchart for duplicating punched cards, suppose you want to change the flowchart so that the computer skips any blank cards and duplicates only those cards with some information punched in them. Since the computer must now make a decision about each card, the decision block will be needed in the flowchart, which will be changed to appear as that shown in Fig. 2-8.

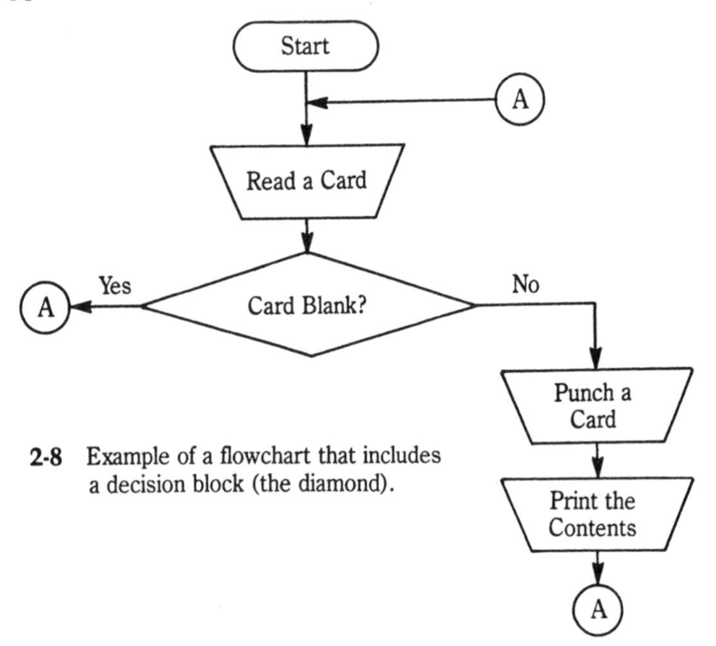

2-8 Example of a flowchart that includes a decision block (the diamond).

Except for the decision block, the flowchart shown here is the same as the earlier one. Begin in the *start* oval at the top and then go to the block marked *read a card*. Will the program now go to the decision block labeled *card blank*? If the answer is yes, go left to the connection circle A and back to the top to read the next card. Only if the card is not blank does the program go right and actually punch a duplicate card and print its contents.

The flowchart used as an illustration here is a very simple one using only input and output devices and doing no calculations. Most programs and flowcharts are not this simple, and this one is used to represent how a flowchart graphically portrays what a computer program actually does.

The field of microcomputers uses many different types of diagrams. There are flowcharts, Venn diagrams, etc. These examples deal mostly with the software portion of this industry. From a purely electronic standpoint, functional diagrams abound and are usually more numerous than the schematic diagrams. From an understanding standpoint, block diagrams are adequate to display all machine functions, but repair necessitates well-defined schematic drawings. Computers take advantage of the latest state-of-the-art developments in electronic components and are relatively simple from this standpoint, especially when you consider all they can do. However, from a pure electronics standpoint and as far as schematic diagrams are concerned, they are highly complex and many pages of schematics are required to represent even the simpler units.

Summary

Block diagramming is an excellent basic form of understanding the functioning of electronic circuits. This type of diagramming is very easy to do from a drafting standpoint and usually requires only a straightedge. Most block diagrams can be drawn in a relatively short period of time when compared with their schematic equivalents.

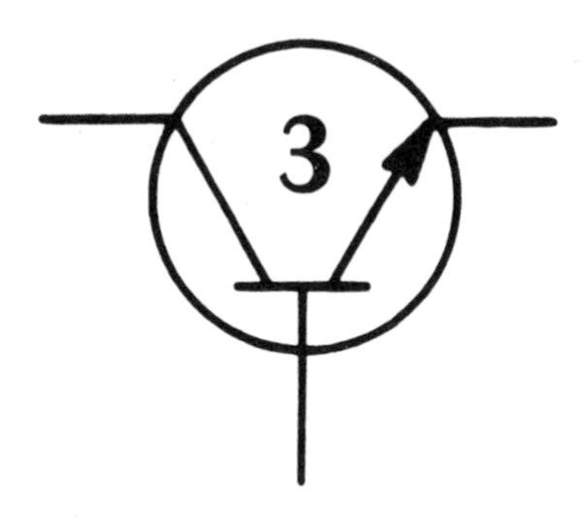

Schematic symbols

ON A TRUE ROAD MAP, DIFFERENT FIGURES ARE USED TO illustrate towns, cities, secondary roads, primary roads, airports, railroad tracks, etc. The same applies to schematic drawings; different symbols are used to indicate conductors, resistors, capacitors, solid-state components, and other electronic parts. Every electronic component manufactured today has an equivalent schematic symbol.

As new types of components come out, which are completely different from all previous ones, a new schematic symbol is derived for each. These changes don't happen very often, even in these days of rapid advancement in the electronics field. Often, a new type of component is a modification of one that already exists. Therefore, the schematic symbol can also be a slight modification of the one used to indicate the preexisting component.

This chapter explains the most common symbols one by one. Refer to Appendix A for a complete alphabetic listing of circuit symbols.

Resistors

Resistors are electronic components. They are aptly named because they resist the flow of electrical current. The value of resistors is measured in *ohms* and typical components might be rated at less than one ohm or more than several million ohms. However, regardless of the resistance value, all resistors are schematically indicated by the symbol shown in Fig. 3-1. Some resistors have the ability to change value. These will be indicated by a slight modification of the symbol shown. The resistor symbol is composed of three full upside down triangles and a half triangle on each end connected to lines which are horizontal, as in the example shown. This is the most universally accepted symbol for the resistor.

3-1 Standard schematic symbol for a resistor.

The two horizontal lines are actually indicators of the leads or conductors that exit from both sides of the resistor. Sometimes the resistor contacts are not wire leads but are in fact metal terminals. Figure 3-2 shows a pictorial drawing of a carbon resistor with leads on either side. This is basically what the component will look like when you purchase it. Figure 3-3 shows pictorial drawings of several other types of resistors that do not contain the basic end leads. However, all resistors shown pictorially here will be indicated schematically by the basic resistor symbol of Fig. 3-1.

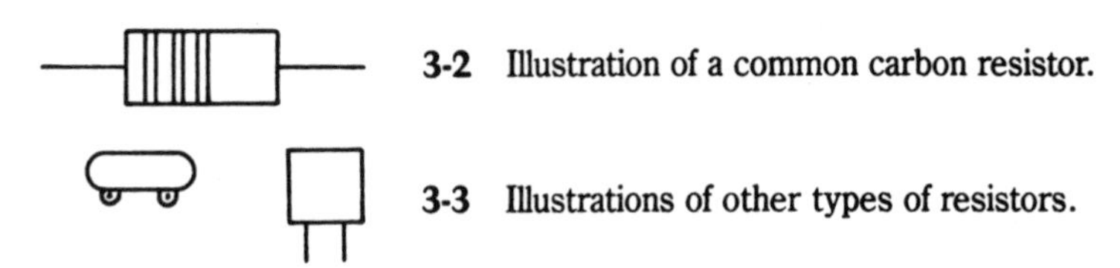

3-2 Illustration of a common carbon resistor.

3-3 Illustrations of other types of resistors.

Variable resistors are those that have the ability to change resistance through a slide tap or a vernier control method. Schematic symbols are used to indicate a specific device and also to serve as an indication of a specific function. The variable resistor is usually set to one value and it remains at this point until manually changed. The electronic circuit still sees this component as a lumped resistance. However, when a variable resistor is required for the proper functioning of a specific circuit, it is necessary to indicate to the person who is building it from a schematic drawing that this must be a variable-type component.

The schematic symbol shown in Fig. 3-4 is that of a *variable resistor*. Again, this is the most standardly recognized symbol. However, others have crept into schematic drawings over the years and they might look more like the example shown in Fig. 3-5. Notice that both examples use the standard resistor configuration and indicate that it is a variable type by using an arrow symbol in conjunction with the triangles. In schematic drawings, an arrow is often used to indicate variable properties of a component—but not always, so don't assume too much at this point. Most types of transistors, diodes, and other solid-state devices also use an arrow as part of their schematic symbols. These arrows in no way indicate any variable properties. A true variable resistor has only two contact points or leads, as indicated by the schematic drawing. Figure 3-6 shows a pictorial example of a variable resistor—usually a wire-wound unit that has been manufactured so that some of the resistance wire is exposed on the surface of the component. A sliding metallic collar is wound around the body of the resistor and can be adjusted to intercept a different turn of resistance wire. This collar is also attached by a flexible conductor to one of the two resistor leads. The collar effectively shorts out resistance turns. As it is advanced toward the opposite resistor lead, the ohmic value of the component decreases.

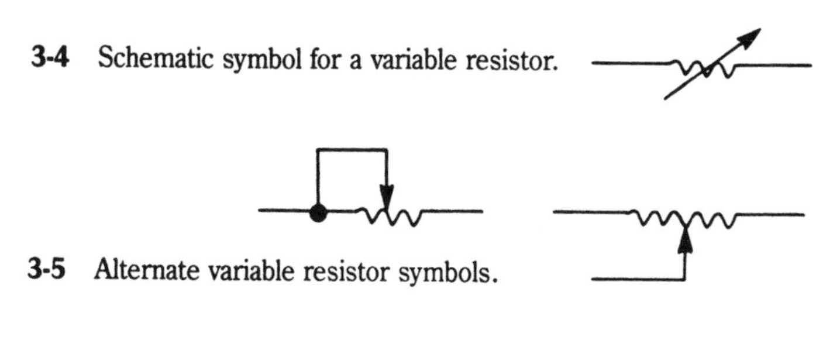

3-4 Schematic symbol for a variable resistor.

3-5 Alternate variable resistor symbols.

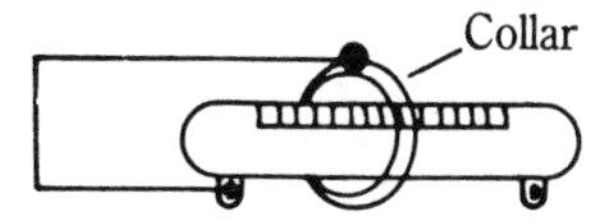

3-6 Pictorial drawing of an earlier variable resistor.

Figure 3-7 shows the schematic drawing for a *rheostat* or *potentiometer variable-resistance control*. Notice that this symbol looks very much like the variable resistor equivalent, but has three discrete contact points. Using the rheostat control, the portion of the circuit that comes off the arrow lead can be varied in resistance to two circuit points, each connected to the two remaining control leads. Figure 3-8 shows a drawing of such a control.

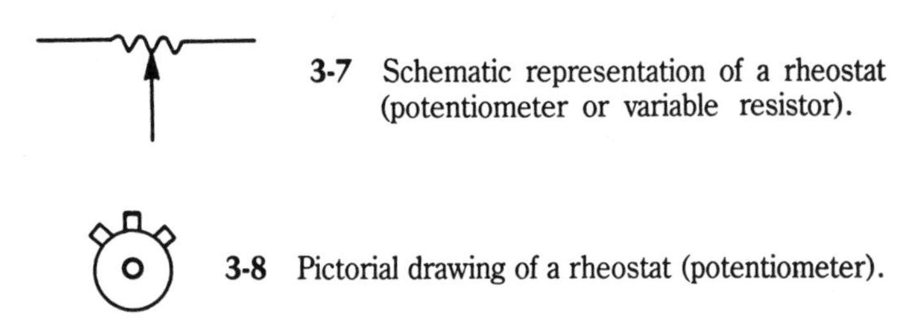

3-7 Schematic representation of a rheostat (potentiometer or variable resistor).

3-8 Pictorial drawing of a rheostat (potentiometer).

Rheostats and most variable resistors are closely related. As a matter of fact, the variable resistor discussed earlier can be changed into a rheostat by simply severing the lead between the collar and one end lead. Now, the collar can be used as the third or variable contact and a rheostat is formed. Likewise, a rheostat can be turned into a two-lead variable resistor by simply shorting out the variable lead with one on either end. Rheostats are often used as variable resistors in this manner.

Resistors of all types are often the most numerous elements in many electronic circuits. It has already been stated that resistors might carry a component value ranging from a fraction of an ohm to several million ohms. The schematic symbol proper in no way gives any indication of the value of the resistor. It is used only to indicate that a resistor is in the circuit. The actual value of the component might be written alongside the schematic drawing, but it might also be given in a separate components table, which is referenced by an alphabetic/numeric designation printed next to the schematic symbol. Appendix B lists the resistor color codes that are used to determine the value of the actual resistor.

Capacitors

Capacitors are electronic components that have the ability to block direct current while passing alternating current. They also can be used to store power. The basic unit of capacitance is the *farad*. This is a tremendously large quantity, and most practical components will be rated in microfarads or in picofarads. A *microfarad* is equivalent to $^{1}/_{1,000,000}$ of a farad, and a *picofarad* is $^{1}/_{1,000,000}$ of a microfarad. Next to resistors, capacitors are the types of components in many electronic circuits.

Figure 3-9 shows the schematic symbol of the basic fixed capacitor. However, it might be seen in other less-approved forms, such as those in Fig. 3-10. Unlike resistors, there are many, many different types of capacitors and some units are very different from others. Some are nonpolarized devices, and others contain a positive and a negative terminal. Most types contain only two leads.

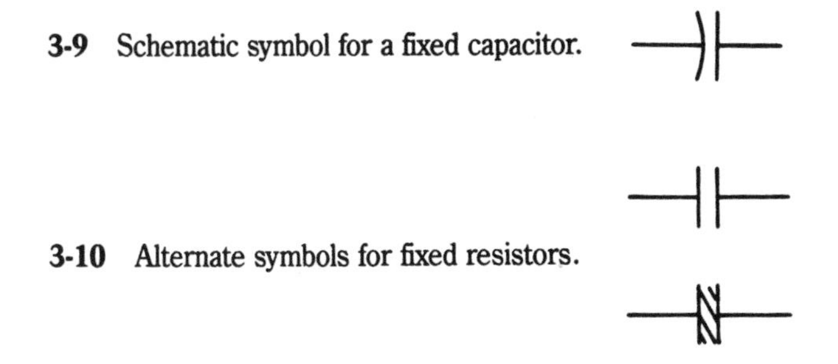

3-9 Schematic symbol for a fixed capacitor.

3-10 Alternate symbols for fixed resistors.

The basic capacitor symbol consists of a vertical line followed by a space and then a half-moon symbol. Horizontal lines connect to the centers of the vertical line in the half moon to indicate the component leads. The symbol discussed indicates a *nonpolarized capacitor*, which might be made from ceramic, mica, mylar, or some other material. The material designation here indicates the insulation that is used between the two major parts of the component. As the symbol might indicate, a capacitor is simply two tiny sheets of conductive material that have been placed close to each other.

Figure 3-11 shows the schematic symbol for a ***polarized*** or ***electrolytic capacitor***. Notice that this symbol is nearly identical to the last one, but a plus (+) sign has been added to the vertical line side. This indicates that the positive terminal of the component will be connected to the remainder of the circuit, as indicated in the schematic drawing. Occasionally, a negative (−) symbol will also be drawn on the opposite side (over the half-moon lead), but this does not follow standard schematic form. When the positive symbol is seen, it identifies the component as an electrolytic capacitor that must be connected to the remainder of the circuit in observance of correct polarity, i.e., the positive capacitor electrode must be connected to the positive voltage lead of the remaining circuit. Electrolytic capacitors themselves contain case markings that indicate the positive lead.

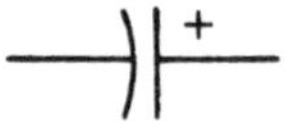

+

3-11 Schematic symbol for an electrolytic capacitor.

To this point, all capacitors and their symbols have been of fixed design. In other words, the components specified do not have the provisions for changing their capacitance value, which is fixed at the time of manufacture. Some capacitors do have the ability to change value. These are usually called ***variable capacitors***, but some specialized types might be known as ***trimmers*** and/or ***padders***.

Figure 3-12 shows the basic symbol for a variable capacitor. Again, an arrow is used to indicate the variable property and is drawn horizontally and through the fixed capacitor symbol. Figure 3-13 shows other ways of indicating this same component, although these are rarely used. Most of the time, the standard symbol shown in Fig. 3-12 will indicate a variable capacitance, regardless of the exact construction of the component. An air variable capacitor is used to tune many AM radios. This component still consists of two basic plates, but their proximity to each other can be changed. The example shown has many interlaced plates, but every other one is connected to the first, forming two distinct contact points.

3-12 Standard symbol for a variable capacitor.

3-13 Alternate, and rare, symbols for variable capacitors.

Variable capacitors are always nonpolarized The electrolytic capacitor is the only type that carries a polarity designation. Many circuits contain a fair sampling of all three types of capacitors discussed here. Sometimes, two separate variable capacitors will be connected together or ganged in a circuit. This means that two or more units are used to control two or more electronic circuits, but both components are varied simultaneously by tying the rotors of the two units together. The rotating plates in a capacitor are referred to as the *rotors* and the fixed plates are the *stators*. Figure 3-14 shows the schematic symbol for two variable capacitors that are ganged together. This simple alteration of the basic variable capacitor symbols involves drawing a dotted line beneath each and connecting it with a horizontal dotted line.

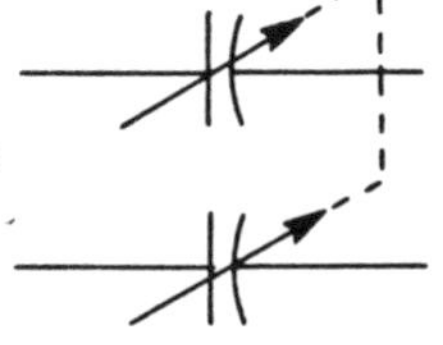

3-14 Schematic symbol for two variable capacitors that are ganged together.

As is the case with most electronic components, the schematic symbol for the capacitor only serves to identify it and to show whether it is fixed, variable, or polarized (electrolytic). The component value might be written alongside or it might be given a letter and number designation for reference to a components list.

Inductors

A basic *inductor* is simply a coiled wire used to introduce inductance into a circuit. *Inductance* is the property that opposes change in existing current and is present only when the current is actually changing. *Coils* or inductors can be tiny or very large, depending upon the inductance value of the component. The basic unit of inductance is the *henry*, a very large electrical quantity. Therefore, most practical inductors are rated in *millihenrys* ($^{1}/_{1,000}$ of a henry) or *microhenrys* ($^{1}/_{1,000,000}$ of a henry).

Figure 3-15 shows the basic schematic symbol for an *air-wound coil*. It consists of a single line that has been used to form five loops. Two leads are designated by the straight lines that eventually curve into the coil at the top and bottom of the symbol. An air-wound coil is one that specifically has no internal core between the windings. However, in practice, a nonconductive and noninductive form such as plastic, mica, or some other insulator is used as a support for the turns. This style is still an air-wound coil, however.

Figure 3-16 shows the schematic symbol for a type of *variable air-wound inductor*. This component is more accurately known as a *tapped-coil* or *tapping arrangement*. Whereas the fixed coil had only two leads, the tapped coil can have three or more. When a coil is tapped, separate conductors are attached to one or more of the turns to offer an alternate connection point. Maximum inductance is obtained from connecting the coils to the circuit at either end. A tapped arrangement allows for the selection of an input or output point that offers lower inductance. Some coils are fitted with a sliding contact that can be continuously advanced throughout

3-15 Standard symbol for an air-wound coil or inductor.

3-16 Schematic symbol for a tapped inductor.

the entire coil. This sliding contact allows continuous adjustment of the inductance value, rather than having a select fixed point with the tapping arrangement. A ***continuously variable coil*** is often indicated by the symbols shown in Fig. 3-17. This symbol indicates the component is capable of being continuously adjusted from a maximum inductance value (determined by the physical size of the coil) to minimum value. Figure 3-18 shows an example of a fixed, tapped, and continuously adjustable air-wound coil.

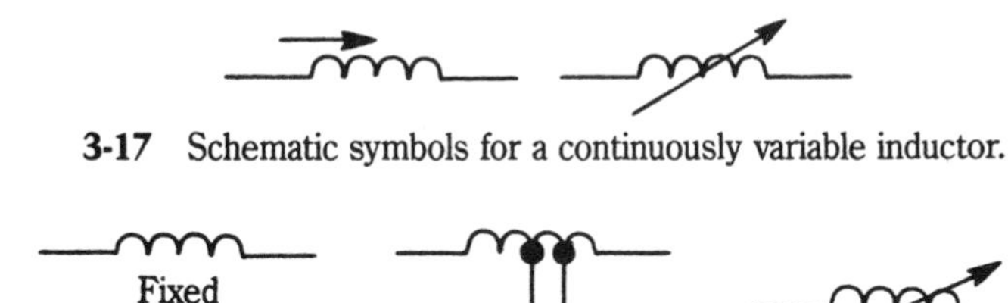

3-17 Schematic symbols for a continuously variable inductor.

3-18 Fixed, tapped, and adjustable air-wound coils.

Not all coils are of the air-wound variety. *Chokes* and other types of coils that are used for low-frequency applications might consist of a coiled conductor that has been wound around an iron core. Here, the iron material replaces the previous empty or air core. For example, 60-hertz chokes might closely resemble an ac power transformer (discussed later) and contain a single coil wound around a circular iron form.

Figure 3-19 shows the schematic symbol for an iron-core inductor. Notice that it is the basic fixed coil discussed earlier, which is immediately followed by two close-spaced vertical lines that run for its entire length. Sometimes the iron-core inductor is drawn as shown in Fig. 3-20. Here, the vertical lines are placed inside the coil turns in the symbol. This is not the approved method of indicating an iron-core inductor, however, it is encountered quite frequently. Some iron-core inductors might also contain taps for sampling different inductance values and some might even be continuously adjustable. The equivalent schematic symbols for these types of components are shown in Fig. 3-21.

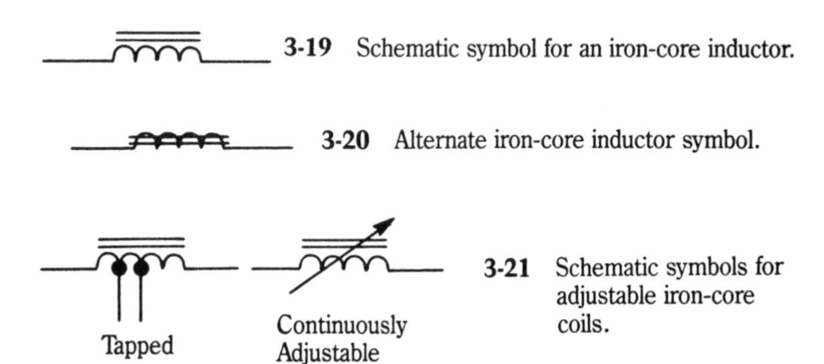

3-19 Schematic symbol for an iron-core inductor.

3-20 Alternate iron-core inductor symbol.

3-21 Schematic symbols for adjustable iron-core coils.

At higher frequencies, iron cores are far too inefficient to be used in inductors. Especially at rf frequencies, a special core is needed and is most often composed of iron material that has been shattered into many tiny fragments, each of which is insulated. After this process has been completed, the particles are greatly compressed to form what appears to be a solid core. The material is called *ferrous iron* or *ferrite* and is known as a *powdered-iron core*. Figure 3-22 shows the schematic symbol for a powdered-iron-core inductor. Notice that the symbol is nearly identical to the iron-core inductor, except that the two vertical lines are broken at several different points. These types of components can also be tapped or made continuously variable. The schematic symbol will be identical to those discussed earlier for variable-iron-core inductors, except that the broken vertical lines will remain.

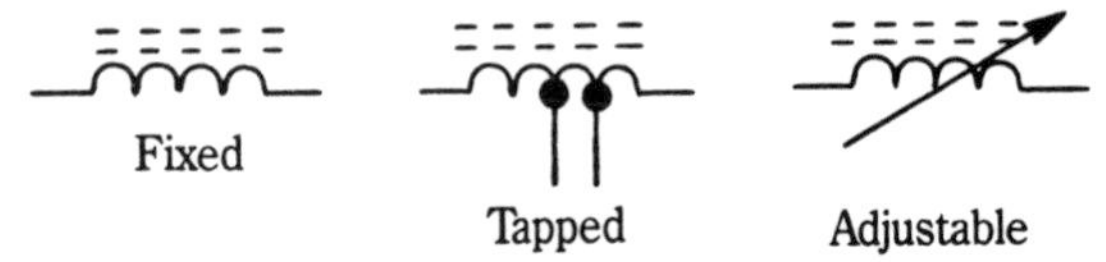

3-22 Schematic symbol for a powered iron-core inductor.

Transformers

Transformers are closely related to inductors and are made when the turns of two or more coils are interspersed. Figure 3-23 shows a basic air-core transformer, which consists of two air-core coils drawn back-to-back. A transformer is a device

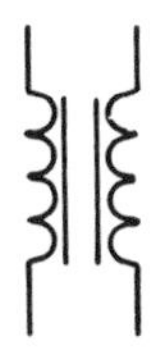

3-23 Schematic symbol for an air-core transformer.

with the ability to transform electric energy from one circuit to another at the same frequency. Since transformers are made by combining inductors, the schematic symbols are very similar. Figure 3-24 shows other types of transformers that contain iron cores, powdered-iron cores that are variable, tapped, etc. Notice that many of the additional symbol lines and indicators are identical to those used in inductor symbology.

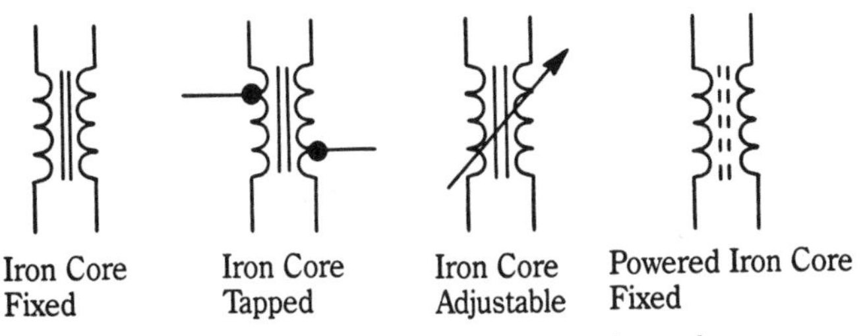

3-24 Schematic symbols for other types of transformers.

Switches

A *switch* is a device, mechanical or electrical, that completes or breaks the path of current. Additionally, a switch can be used to allow current to pass through different circuit elements. Figure 3-25 shows the schematic symbol for a *single-pole/single-throw (spst) switch*. The spst variety is capable of making or breaking a contact at only one point in a circuit. Notice that the arrow is used here to indicate variability. This type of switch is a two-position device (on-off/make-break). Figure 3-26 shows a different type of switch, designated as a *single-pole/double-throw*

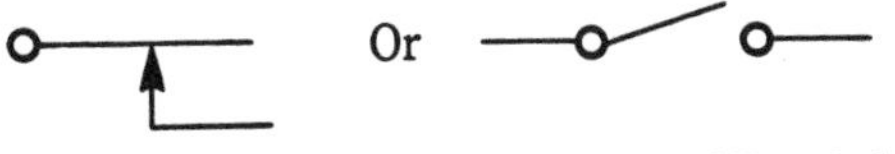

3-25 Schematic symbol for an SPST switch.

3-26 Schematic symbol for an SPDT switch.

(spdt) variety. For basic understanding, the pole is the point of contact at the base of the arrow. The throw(s) is the contact point to which the arrow can be attached. The spdt switch contains one pole contact and two throw positions; the input to the pole can be switched to either the left-hand or right-hand circuit point.

Some switches contain two or more poles. An example of a *double-pole/single-throw (dpst) switch* and a *double-pole/double-throw (dpdt) switch* is shown in Fig. 3-27. Some switches have even more elements. The one shown in Fig. 3-28 has five poles, each of which can be switched to two separate positions. Therefore, this is a *five-pole/double-throw* arrangement.

This last designation can actually be covered under the heading of *multi-contact switches*. This category takes in most switches that have more than two poles or two throw positions. For example, a *rotary switch* is a device that has a single pole

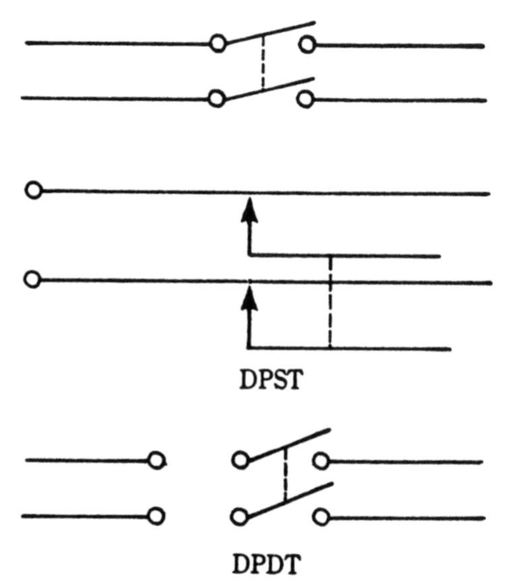

3-27 Schematic symbols for a DPST and a DPDT switch.

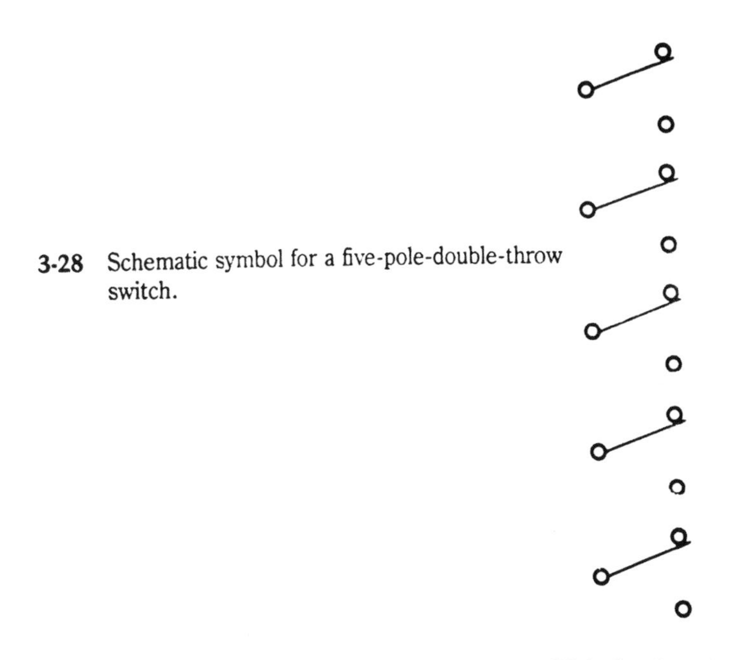

3-28 Schematic symbol for a five-pole-double-throw switch.

and sometimes ten or more throw positions. This basic category of switch is shown in Fig. 3-29. The arrow still indicates the pole contact.

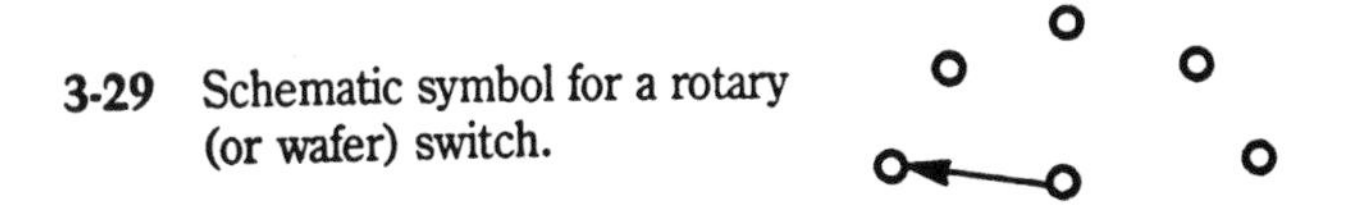

3-29 Schematic symbol for a rotary (or wafer) switch.

Occasionally, rotary switches will be ganged together, much like variable capacitors were ganged together in a previous discussion. Figure 3-30 shows the schematic of an arrangement that uses two rotary switches. Notice that again, the dotted line is used to indicate the ganged setup.

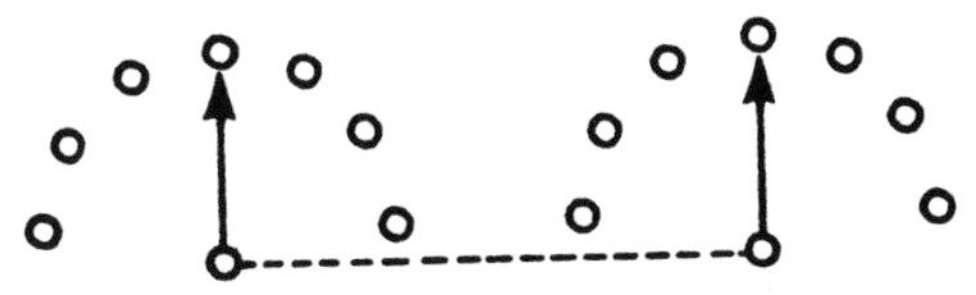

3-30 Schematic symbol for two rotary switches ganged together.

In each case, the switch contact point (pole or throw) is represented by a small circle. The variable element or pole is indicated by an arrow. The symbols shown are all standard. Because of the simplicity of this symbol, variations are seldom seen. A special type of switch, however, is used in some radio communications work called a *code key*, *Morse key*, or simply *key*. This device is used to make and break an electrical contact for the purpose of sending code. A key is simply an spst switch that contains a spring allowing it to return to the off-position automatically. The symbol for a key is shown in Fig. 3-31.

3-31 Schematic symbol for a code key (hand key).

Conductors and cables

Throughout this discussion, a straight line has always been used to indicate a conductor, but most circuits contain a large number of conductors. It often becomes necessary to have them cross over each other or to actually make contact. Figures 3-32 and 3-33 show the standard procedure for indicating what is occurring when two conductors cross. Figure 3-32 shows two conductors that have crossed, but not made contact. This does *not* mean, by any stretch of the imagination, that when building the circuit, the conductors must cross over each other. It simply means that in order to make the schematic drawing, it was necessary to draw one conductor across another to reach various circuit points. Notice that at the junction, a half loop is drawn in one conductor to indicate that no connection has occurred. The loop could just as easily have been drawn in the horizontal line to indicate the same condition.

Figure 3-33 shows an actual wiring connection. Here, the two conductors cross (at right angles in this example), and a black dot is drawn at the junction. This dot indicates the conductors connect at this point. The drawing of conductors is the most-often-abused part of schematic preparation. Occasionally,

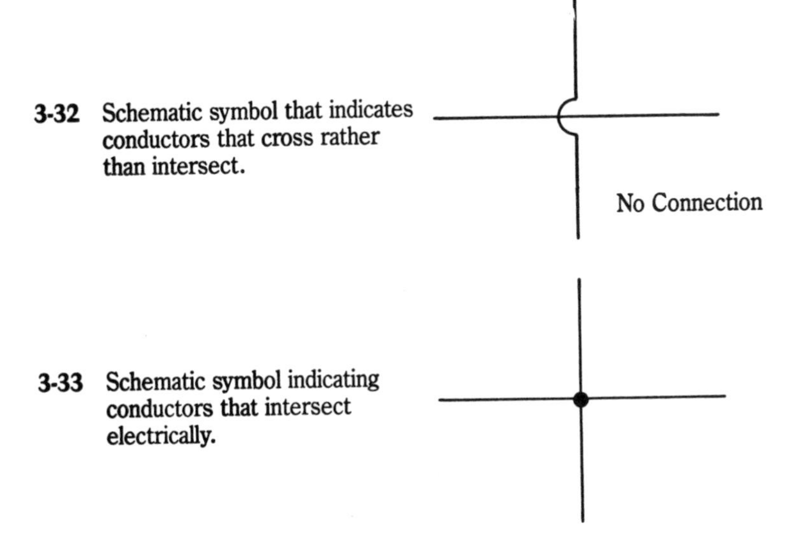

3-32 Schematic symbol that indicates conductors that cross rather than intersect.

3-33 Schematic symbol indicating conductors that intersect electrically.

a nonconnection will be drawn as shown in Fig. 3-34. A connection will still be indicated by a black dot at the junction, but it still leaves you wondering. Sometimes this same arrangement will be used to indicate a connection, and a nonconnect crossover will be indicated by the half loop. When other than standard methods are used, it becomes necessary for the person reading the schematic to decipher just what the artist meant by hunting down a nonconnect example or a connect example to compare the other to.

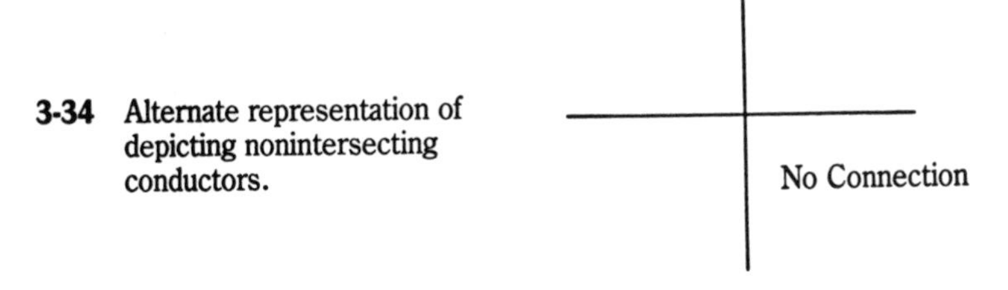

3-34 Alternate representation of depicting nonintersecting conductors.

A *cable* consists of two or more conductors usually contained in the same insulating jacket. Quite often, unshielded cables will not be specifically indicated in a schematic drawing, but are simply shown as two discrete symbol changes. Figure

3-35 shows an example of a *shielded wire*, often used to indicate the use of *coaxial cable* in an electronic circuit, especially one that is operating at radio frequencies. Coax consists of a single conductor surrounded by a metallic shield for the entire length of the cable. An insulator keeps the two conductive elements isolated from each other.

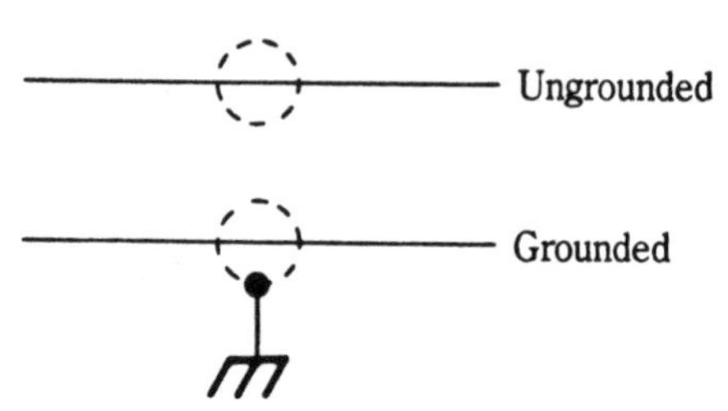

3-35 Schematic symbol for a shielded wire, which can be either grounded or ungrounded.

The symbol for shielded wire is drawn by placing a small circle over the conductor. The circle is attached to a vertical conductor which, in turn, is connected to ground. Here, the ground symbol consists of four tapered lines. Alternately, the ground symbol can be indicated as shown in Fig. 3-36. This is a horizontal line, to the bottom of which are attached three short vertical lines. This symbol is sometimes called a *rake* in electronics jargon. Many persons use the ground and rake symbols interchangeably, but the latter is actually used to indicate a chassis connection, and a chassis might or might not be connected to earth ground.

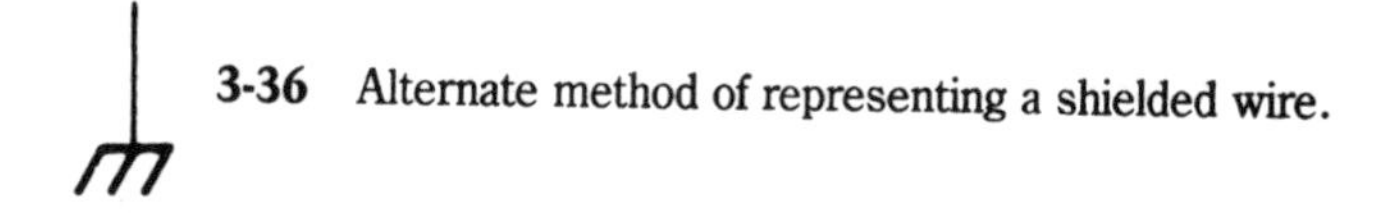

3-36 Alternate method of representing a shielded wire.

Sometimes a shielded cable is necessary for electronic construction. This consists of two or more conductors which are surrounded by a single shield. The schematic symbol for this cable is shown in Fig. 3-37. This symbol is identical to the one for single shielded wire, except that an extra conductor has been added. If the cable contained five conductors, then

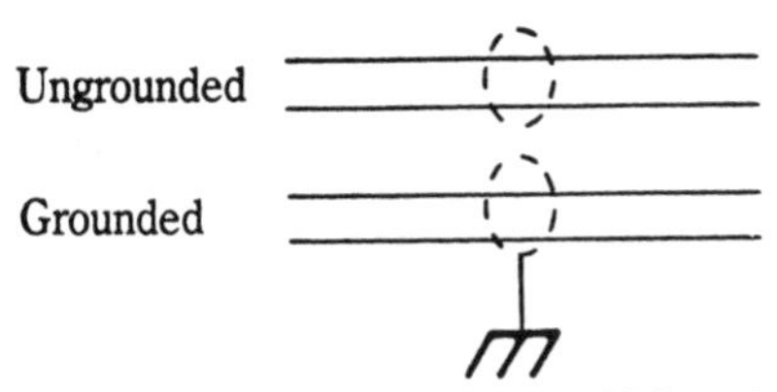

3-37 Schematic symbol for a shielded cable, which again can be grounded or ungrounded.

the circle (indicating the shield) would contain five horizontal lines.

Solid-state components

Solid-state components are numerous and varied and include *transistors*, *diodes*, *thyristors*, *solar cells*, and other specialized devices. Figure 3-38 shows the basic symbol for a diode or rectifier. The symbol is an arrow lead contacting a flat surface that contains another lead. Several ways are shown for indicating a diode, but the most prevalent one is indicated first. Diodes are made of germanium, silicon, or selenium, but they are all drawn as shown. Specialized diodes include *varactor*, *zener*, and *tunnel* types. Schematic symbols for these are shown in Fig. 3-39. Again the basic arrow symbol is used, but in some instances, the flat surface is altered. Each of these special types of diodes is encompassed by a circle. The varactor diode contains a tiny capacitor symbol inside the circle. This diode has the ability to change capacitance, which is the reason for this addition. Often, zener and tunnel diodes will be drawn without the containing circle, and sometimes they will be indicated as a standard diode, but are further defined in the components list.

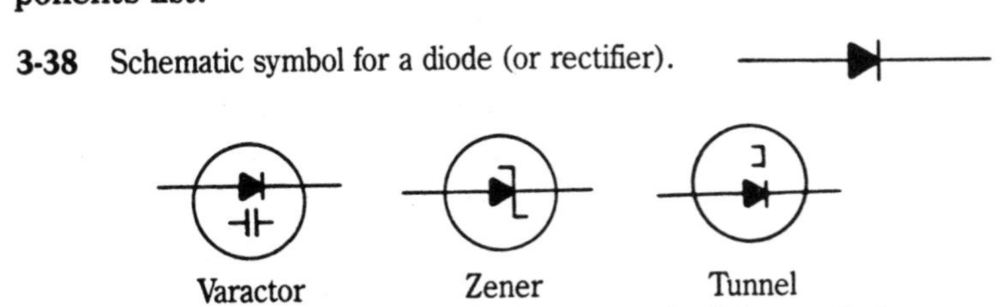

3-38 Schematic symbol for a diode (or rectifier).

3-39 Schematic symbols for other types of solid-state diodes.

A *silicon-controlled rectifier* (SCR) is a special three-element diode, and the schematic symbol is shown in Fig. 3-40. Again, the circle is used and the third element (the gate) is indicated by a diagonal line connected to the component-lead symbol. In all cases, the lead that is attached to the arrow is the *anode* of the device and the one connected to the flat surface is the *cathode*.

3-40 Schematic symbol for a silicon-controlled rectifier (SCR).

Figure 3-41 shows the basic schematic symbols for ***bipolar transistors***. The ***pnp*** type is shown first, followed by the ***npn*** variety. The only distinction between the two is the direction the arrow is pointing. In the pnp, the arrow points into the flat line or base electrode. In the npn, the direction of the arrow is reversed. Occasionally, the circle that encompasses the base, emitter, and collector leads is omitted, but this is not standard practice.

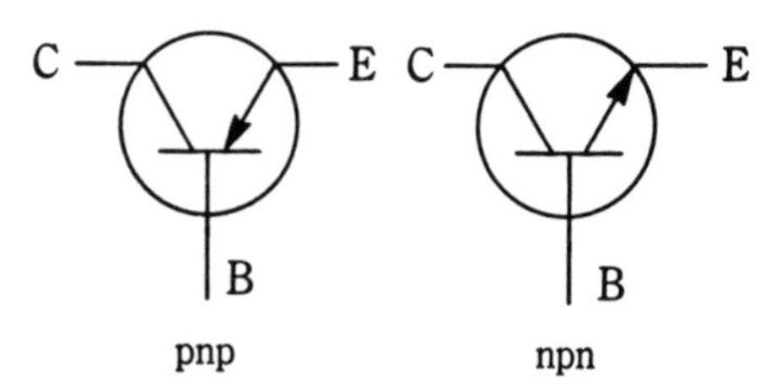

3-41 Basic symbols for bipolar transistors.

There are many other types of transistors (Fig. 3-42). In every case, an arrow is used in conjunction with a horizontal line and the entire work surrounded by a circle. Transistors can be made from silicon or germanium, but the schematic symbol by itself will not indicate the type of semiconductor material used.

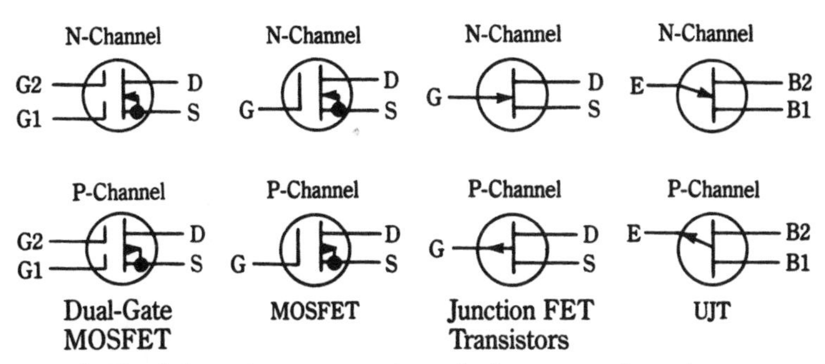

3-42 Schematic representations of other types of transistors.

Vacuum tubes

Although *vacuum tubes* are not used in electronic construction nearly as often as they were two decades ago, many designs that use these devices still exist. Drawing the symbol for a vacuum tube consists of adding the symbols for the tube elements together in such a manner as to symbolize the type of tube being displayed. Figure 3-43 shows the schematic symbols for the various types of tube elements commonly used in schematic drawings. Some of these are used in displaying every type of vacuum tube.

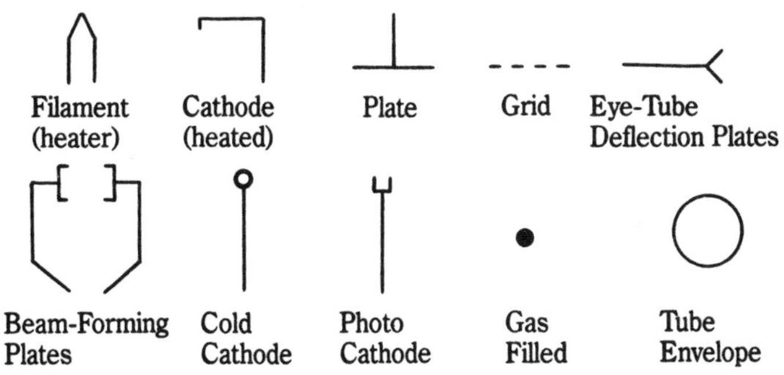

3-43 Schematic symbols for various types of tube elements.

Figure 3-44 shows the schematic symbol for a *diode vacuum tube*. This two-element device contains a plate and a cathode. A filament is used to heat the cathode, so in reality, three element symbols are actually used to display this type of

3-44 Schematic symbol for a diode vacuum tube.

device. In electronics terminology, the filament is often considered to be part of the cathode. Notice that the symbols for some of the tube elements from the previous chart have been used to construct this tube symbol. All tube elements are surrounded by a circle. Occasionally, the circle will be omitted from some schematic drawings, but this is not an approved practice.

Figure 3-45 shows a *triode vacuum tube*, which consists of the same elements as the diode previously discussed, with the addition of a dotted line to indicate the grid. *Tetrode vacuum tubes* have two grids, so to draw this latter device, an additional dotted line is included (Fig. 3-46). Figure 3-47 shows other types of tube symbols that will commonly be encountered.

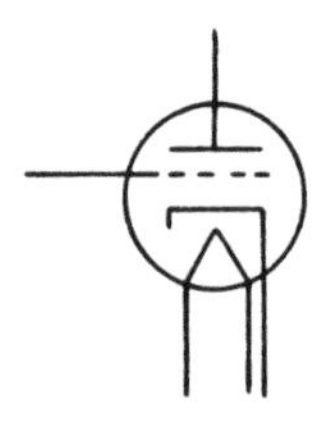

3-45 Schematic symbol for a triode vacuum tube.

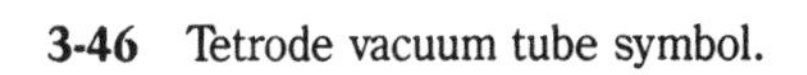

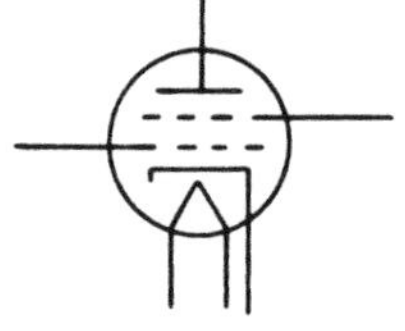

3-46 Tetrode vacuum tube symbol.

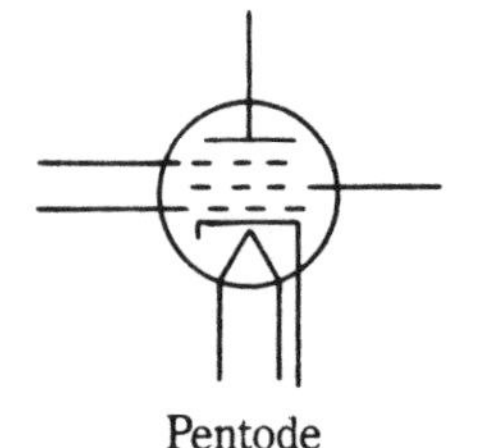

3-47 Pentode vacuum tube symbol.

Some vacuum tubes actually consist of two tubes housed in a single envelope, known as *dual tubes* or *dual pentodes, dual tetrodes, dual triodes*, etc. Figure 3-48 shows how some of these tubes are represented schematically.

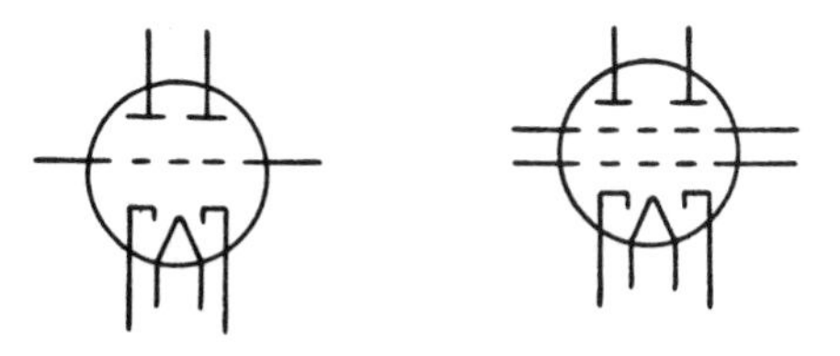

3-48 Schematic symbols identifying two types of dual vacuum tubes.

A *cathode-ray tube* is a special vacuum-tube device and is indicated by the schematic symbol shown in Fig. 3-49. The shape of the envelope is indicated pictorially, but the internal tube elements still follow the previous patterns, in that plates, grids, cathodes, and filaments are shown.

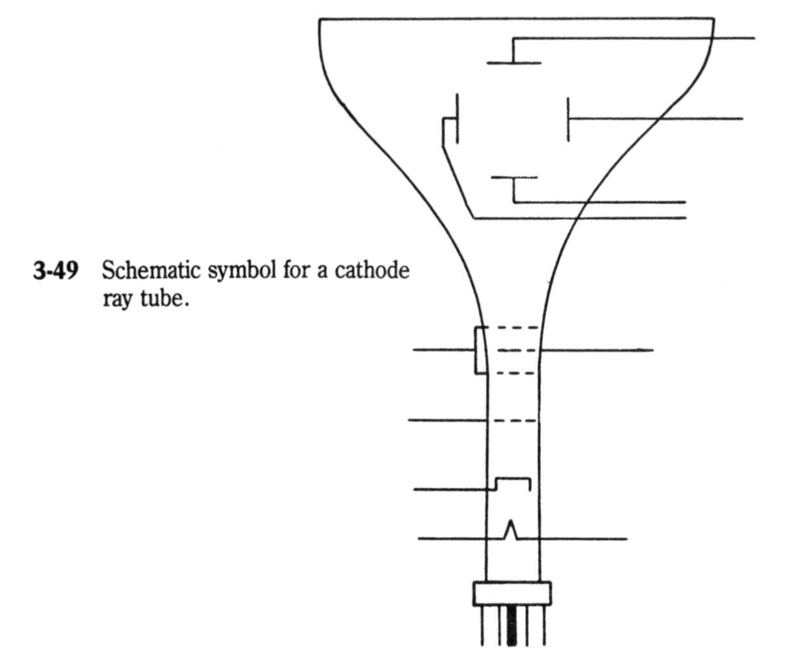

3-49 Schematic symbol for a cathode ray tube.

Batteries

A *battery* is often used as a power source for electronic circuits and must also be indicated schematically to aid the builder and serviceman. Figure 3-50 shows the schematic symbol for a *single-cell battery*. A single-cell component such as this will usually have an output of approximately 1.5 volts dc. Batteries with higher voltage outputs are usually composed of several single cells, and the schematic representation for a *multi-cell* design takes this into account, as shown in Fig. 3-51. The multi-cell symbol is simply a number of single-cell symbols combined in series. If a circuit calls for the use of three single-cell batteries in a series connection, the symbol would actually be composed of three single-cell symbols in series (Fig. 3-52). The difference between a series connection of three single-cell batteries and the symbol for a multi-cell battery can be clearly seen.

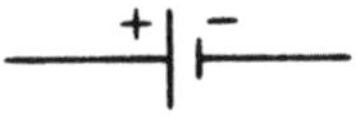

+ −

3-50 Schematic symbol for a single-cell battery.

+ −

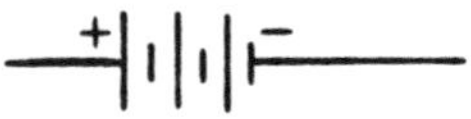

3-51 Schematic symbol for a multicell battery.

+ + + −

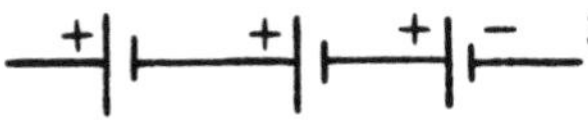

3-52 Schematic symbol for three single-cell batteries connected in series.

Notice that a positive and a negative polarity sign are included with every battery schematic symbol. Sometimes these are omitted, so it must be remembered that the taller vertical line is always the positive connection and the shorter one is negative. Standard practice calls for these symbols to always be included when drawing schematics. However, some artists will neglect this and the person reading the schematics will have to decipher the intent of the entire design.

Other schematic symbols

There are many other schematic symbols that are encountered in common practice. The ones discussed thus far are very common and will be used quite often. Appendix A shows a complete table of schematic symbols that are commonly used in electronics applications. In addition to the ones already discussed, you will see symbols for *jacks* and *plugs*, *solenoids*, *piezoelectric crystals*, *lamps*, *microphones*, *meters*, *antennas*, and many other electronic components. It might seem like quite a chore to memorize all of these symbols, but their usage and correct identification will come with time. The best way to begin the memorization process is to read simple schematics and refer to this chart whenever a symbol crops up that you cannot identify. After an hour or so, you should be able to move on to more complex schematics, again referencing the unknown symbols. After a few weekends of practice, you should be thoroughly familiar with most electronic symbols used in schematic representations.

Summary

Schematic symbols are a system unto themselves, but most are based upon the actual physical or working structure of the component or device they represent. Schematic symbols are often presented in groupings, each of which has some relationship to the others. For example, there are many different types of transistors, but all of them are represented in similar manners; minor symbol changes indicate a different type of device, but all can be easily identified as some type of transistor. The same rule applies to vacuum tubes. Nearly every type is represented by a circle that contains a number of tube elements. All resistors are indicated by the triangle pattern, but some will contain an extra symbol portion to represent a special usage. Although schematic symbols are a system unto themselves, this is a logical system, and one that can become a quick study for the individual who pursues this field diligently.

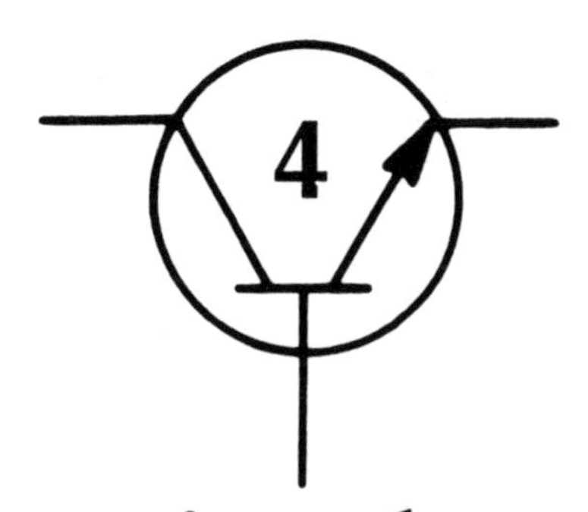

Simple electronic circuits

TWO SCHOOLS OF THOUGHT ARE USED WHEN TEACHING people to read and write schematic drawings. One school feels that the student should learn to read them before the writing process begins. The other school feels that the student should learn to read schematic drawings by writing them. I encompass the best of both worlds by advocating a combination of the two. Certainly, you must learn the basic symbols at the very start. After this is done, you should attempt to read as many schematic drawings as possible. When this gets boring, then switch to making your own drawings. If you rotate back and forth during the study period, you will probably gain a better overall knowledge. For this reason, you should devote half your study time to reading symbols and the other half to writing them.

This chapter deals with reading and writing some extremely simple electronic circuits that will be shown pictorially and then schematically. Using this method, you can actually see the circuit and then see how the schematic representation is drawn from it. Some commercial schematics are produced in this manner. However, in most instances, a circuit will be designed schematically first, then built and tested from the schematic. If a circuit is highly experimental, some bugs will be in the test mockup, which will require some component deletions, substitutions, or modifications. When these changes are made to the test circuit, the results are noted and the schematic is changed accordingly. In the end, the finished and corrected schematic is a product of design theory, actual testing, and modification.

Getting started

Figure 4-1 shows a simple circuit that we have all used at one time or another. Basically, this is a flashlight with the external case removed. The flashlight consists of a battery and an electric bulb. This pictorial representation also shows the conductors, which attach to the light bulb and the battery. The conductors form a current path between the battery and the light bulb. Current flows from the negative terminal of the battery through the bulb element and back to the positive terminal of the battery. In the pictorial representation, the positive and negative terminals are indicated.

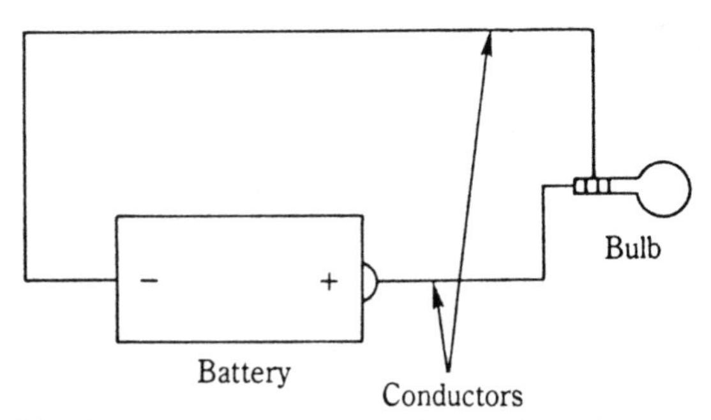

4-1 Pictorial drawing of a flashlight circuit using a single battery.

In order to make a schematic diagram of this simple circuit, it is necessary to know three schematic symbols. These represent the battery, the conductors, and the bulb (Fig. 4-2). Now that we know the symbols that are needed, they can be assembled in a logical manner based upon the appearance of the circuit in the pictorial drawing.

Start first by drawing the battery symbol. The battery can be thought of as the heart of the circuit, since it supplies

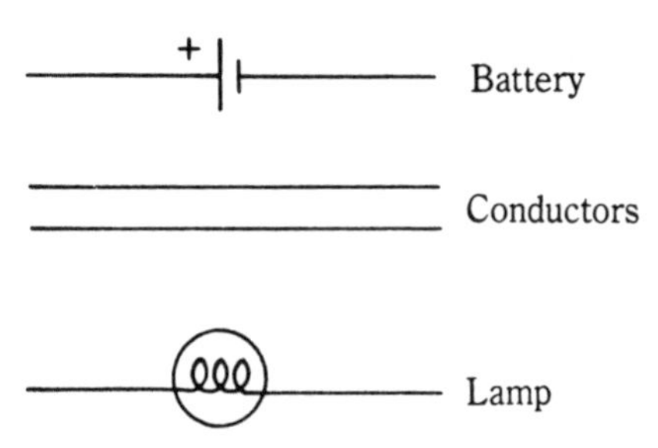

4-2 Schematic symbols for the battery, conductors, and bulb used to make the flashlight circuit in Fig. 4-1.

all power. Next comes the symbol for the light bulb, which can be drawn at any point near the battery. Using this example, try to make the schematic symbols fall in line with the way the pictorial diagram was presented. This layout places the light bulb to the immediate right of the positive battery terminal.

Now that the two major symbols have been drawn, it's simple to use the conductor symbols to hook them together. Notice that the pictorial drawing shows two conductors. Therefore, two are used in the schematic diagram as well.

Figure 4-3 shows the completed schematic drawing, which is the symbolic equivalent of the pictorial drawing we originally worked from. The schematic drawing indicated is by no means the only way this simple circuit can be represented. However, any schematic representation will require the use of the basic symbols outlined. The only changes that can occur involve the positioning of the component symbols on the page. Figure 4-4 shows several different methods of indicating the same circuit schematically. They are all electrically

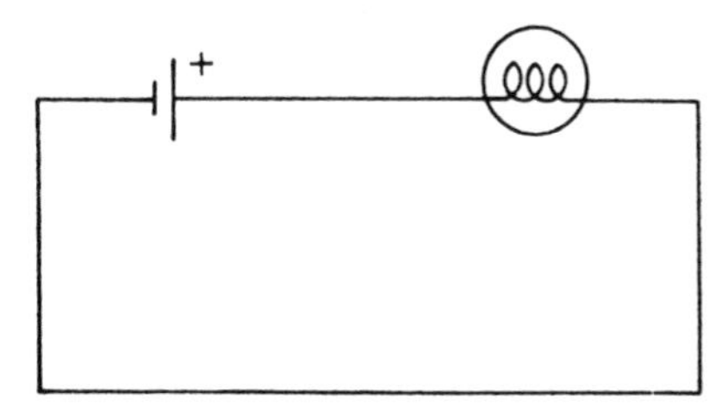

4-3 Schematic diagram of the former circuit.

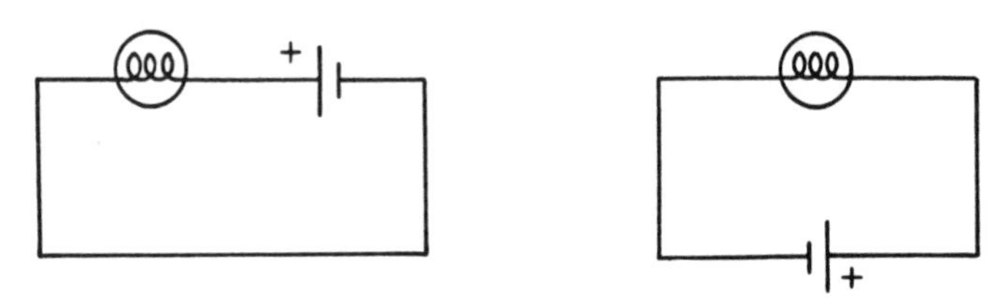

4-4 Different methods of representing the flashlight circuit.

equivalent, but appear different as a result of their relative positions. Notice that in every case the individual symbols are correctly drawn. This circuit uses a single-cell battery, so the single-cell symbol is incorporated, along with its positive and negative polarity markings.

Let's alter this circuit a bit to gain more proficiency in reading and writing schematic drawings. Figure 4-5 shows the same basic circuit, but an additional battery and a switch have been added. This configuration is standard for all flashlights sold in this country. By examining the pictorial drawing, you can see that any schematic representation must contain two battery symbols, the conductors, the light bulb, and the on/off switch. The switch is the only new symbol to be added to this circuit—the single-cell battery symbol will simply be repeated. Figure 4-6 shows the symbols that will be needed to assemble an accurate schematic drawing of this circuit. Again, the symbols are drawn on paper in the same basic order as the components they represent are wired in the circuit. The

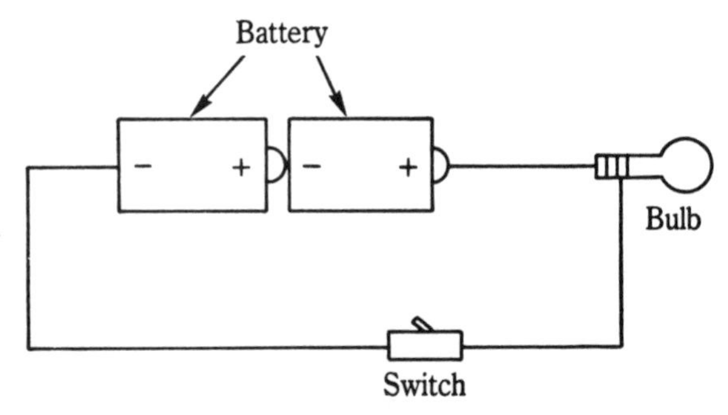

4-5 Pictorial drawing of another flashlight circuit using two batteries in series.

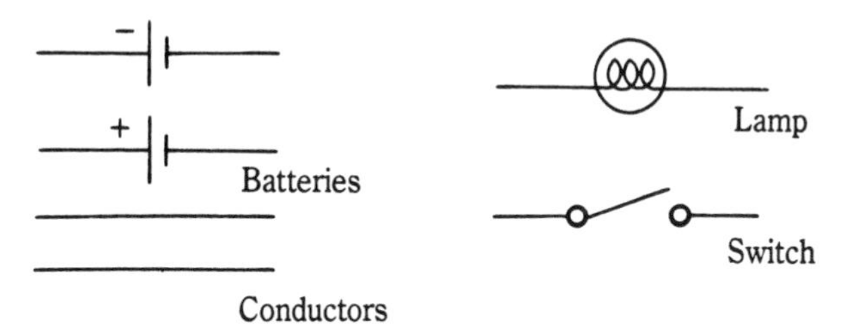

4-6 Schematic symbols that make up the dual-battery flashlight circuit.

resulting schematic is shown in Fig. 4-7. Note that the two single-cell battery symbols are drawn in series, with polarity markings provided for each. Using a series connection, the positive terminal of one battery is connected to the negative terminal of the other. The same two conductors are used from the battery terminals, but a third one is needed to connect the switch to the light bulb. The switch is shown in the off position, which is common for this type of representation. Congratulations! You now know what a common flashlight looks like when represented through schematic symbology.

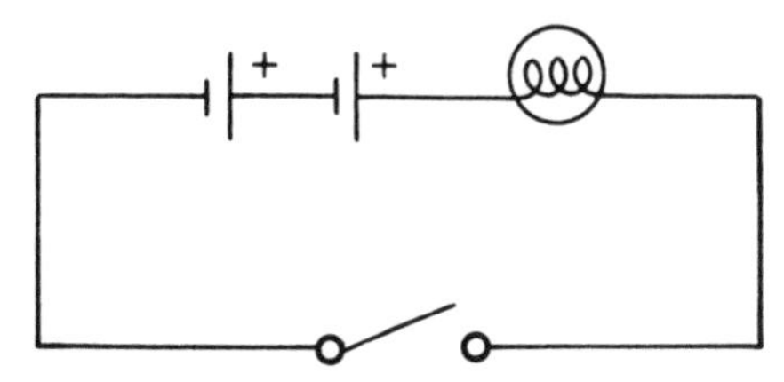

4-7 Schematic diagram of the dual-battery flashlight circuit.

It's time to move on to a slightly more complex circuit. Figure 4-8 shows a device known as a *field-strength meter*. This drawing is a pictorial representation, and the circuit consists of an antenna, a diode, a meter, and a coil. For this discussion, the values of the various components are unimportant. In order to draw this circuit schematically, it is necessary to use the symbols for antenna, diode, meter, coil, and variable resistor (Fig. 4-9). Using the same method as before, you can draw the schematic representation of this simple device in a short period of time by connecting the symbols in the same order as the components they represent in the circuit.

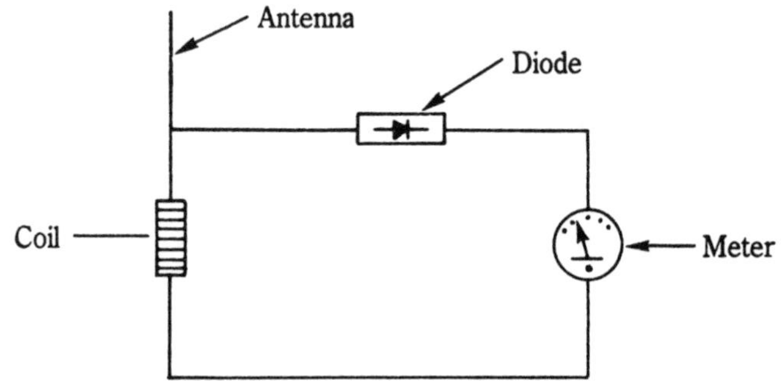

4-8 Pictorial representation of a field-strength meter.

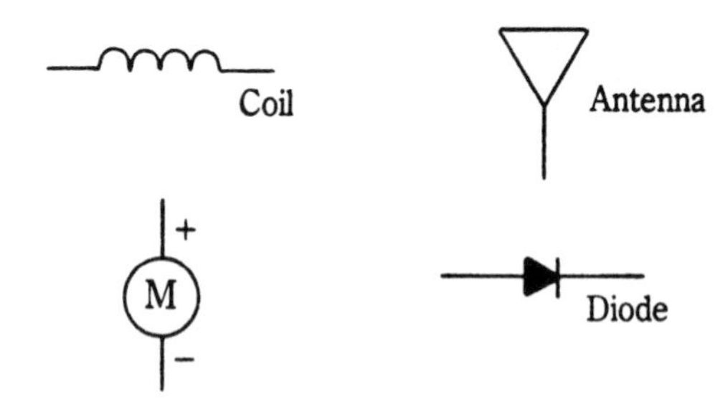

4-9 Schematic symbols used to form the schematic drawing of the field-strength meter.

Figure 4-10 shows how the completed schematic might look. Notice that this drawing only involves a simple substitution of schematic symbols for actual component parts. As before, the parts need not be physically placed in the order shown, but they must be interconnected exactly as indicated or the circuit represented will be incorrect.

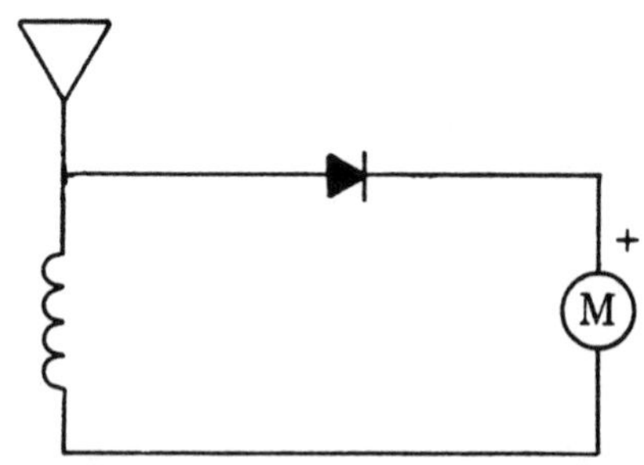

4-10 Completed schematic diagram of the field-strength meter.

By now you should be getting the idea. Previously, schematic drawings have been compared to road maps. A road map is supposed to indicate exactly what the motorist will experience in practice. The schematic drawing does the same thing. We can make further analogies by saying that the highways indicated on the road map that interconnect various towns and cities are very similar to the conductors in a schematic diagram that connect electronic components. The latter are analogous to the towns and cities. A road map allows the motorist to get from one point to another, whereas a schematic diagram provides a definite route for current through the various components in the circuit.

As another example of a simple schematic, a diagram is shown of an ac-derived dc power supply in Fig. 4-11. Reading from the left, an ac male power plug is connected to the primary winding (the one on the left) of the transformer through a fuse. At the secondary winding of the transformer (the one on the right), a diode is connected in series. Following this, an electrolytic capacitor [note the plus (+) sign] is connected in parallel between the output of the rectifier and the bottom lead of the transformer secondary. Also connected in parallel with the capacitor and between the same two circuit points is a fixed resistor. The circuit, which could be completed from this schematic, could be very small or very large, depending upon the voltage and current that are to be delivered. Since dc power supplies have polarized outputs, positive and negative signs are used to indicate the output polarities. Any half-

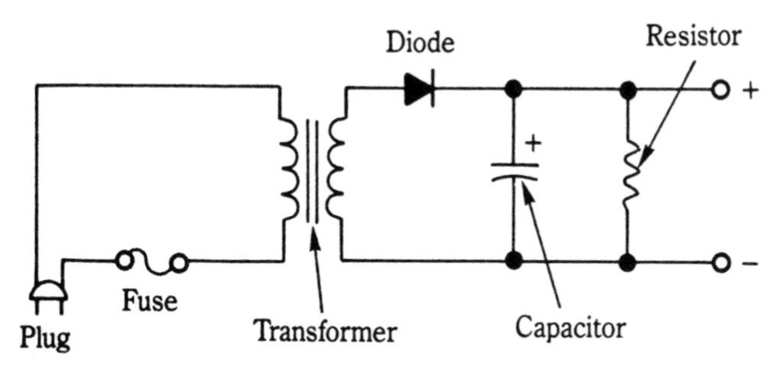

4-11 Schematic diagram of a simple dc power supply.

wave power supply that uses a single diode, capacitor, and resistor will look exactly like this. If the output is to be 5 volts at 1 ampere or 5,000 volts at 50 amperes, the basic schematic drawing will be identical. This is not to say that special features and additions could not be added to either supply and be reflected in the schematic drawing. Rather, two dc power supplies using a very basic design will appear identical regardless of the values of the components used.

Figure 4-12 shows the same schematic drawing as before, but this time each component is given an alphabetic/numeric designation. These references each refer to a components list chart that is also included. Now, you can see that this particular schematic uses a transformer with a 115-volt primary and a 12-volt secondary; a diode rated at 50 peak inverse volts and a forward current of 1 ampere; a 100-microfarad, 50-volt capacitor; and a 10,000-ohm, 1-watt carbon resistor. The fuse is rate at 1/2 ampere.

This schematic is practical and can be used to build a power supply with a peak output of approximately 30 volts dc. Before each component was referenced, however, the schematic had no use in a practical sense, other than to illustrate the basic components of all half-wave power supplies.

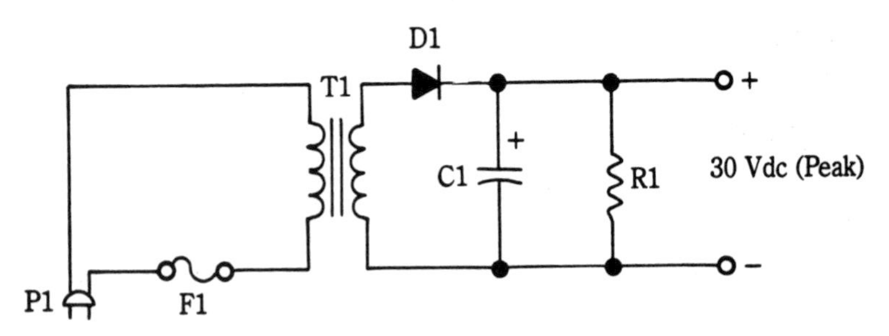

C1 - 100 Microfarad Electrolytic 50 Vdc
D1 - 50 PIV 1 Ampere
F1 - 1/2 amp, 125 Volts
P1 - Male Line Plug
R1 - 10,000 Ω 1 Watt Carbon
T1 - 115 Volt Primary - 12 Volt Secondary 1 Ampere

4-12 Schematic of Fig. 4-11 shown with the appropriate alphabetic/numeric designations.

The letters used to identify each component are more or less standard. Notice that each letter is followed by the number 1. The designation T1, for instance, indicates that the component is a transformer (T) and that it is the first (1) of this type of component referenced. If two transformers were used in this circuit, one would be labeled T1 and the other T2. The numbers simply reference the proper position on the components list and serve no other purpose. The diode is referenced as D1, with D being the standardly used letter for this particular component. Standardization is not universal, however; in some instances, the diode might be labeled SR1. The SR stands for *silicon rectifier*. Some zener diodes might be labeled as ZD1, ZD2, etc. This labeling makes little difference, however, since the component designations are written close to their symbols. If the designation D1 was replaced with SR1, there would be no doubt that the letter was intended for the symbol for a diode.

Using the example shown, it is unnecessary to include a number next to each component designation because only one of each component is used to make up the entire schematic. It is standard practice, however, to always include a letter and a number to prevent any chance of misinterpretation. In highly complex electronic drawings, several hundred components might be used, many of which are from the same family. For instance, if you saw the designation D101, this would indicate that there are at least 101 diodes in the entire circuit and if you want to know the type and value of this particular component, it will be necessary to reference D101 in the components list. Figure 4-13 shows the letter designations for the various types of electronic components. It should be understood, however, that these can vary slightly, depending upon the style and proficiency of the person making the drawing. Learning this chart will be quite simple, as most of the designations used are the first letters of the components they represent. If the component has a complex name, such as silicon-controlled rectifier, the first letters from each of the three name portion is used, i.e., SCR1. In this case, S1 could not be used because S is the designation

ANT	Antenna
B	Battery
C	Capacitor
CB	Circuit Board
CR	Zener Diode (occasionally, any diode)
D	Diode
EP or PH	Earphone
F	Fuse
I	Lamp
IC	Integrated Circuit
J	Receptacle, Jack, Terminal Strip
K	Relay
L	Inductor, choke (usually audio frequency)
LED	Light-emitting diode
M	Meter
N	Neon lamp (rare)
P	Plug
PC	Photocell
Q	Transistor
R	Resistor
RFC	Radio frequency choke
RY	Relay
S	Switch
SPK	Speaker
SR	Selenium rectifier
T	Transformer
U	Integrated Circuit
V	Vacuum tube
VR	Voltage regulator (tube-type usually)
X	Solar cell (rare)
XTAL or Y	Crystal
Z	Circuit assembly (block diagram designation)
ZD	Zener Diode (rare)

4-13 Letter designations for the most common schematic components.

for a switch. A resistor is designated by the letter R. Therefore, a component such as a relay, whose name also starts with the letter R, must have a different designation, i.e., RY1. Once you know a bit about this system of symbols and designations, the whole procedure seems much more logical.

Figure 4-14 shows a schematic drawing and components list for a more complex type of dc power supply. Many components and their basic wiring the identical to the previous power-supply schematic drawing. However, additional parts have been added, and some components have been duplicated. Notice that the letter designations remain the same for

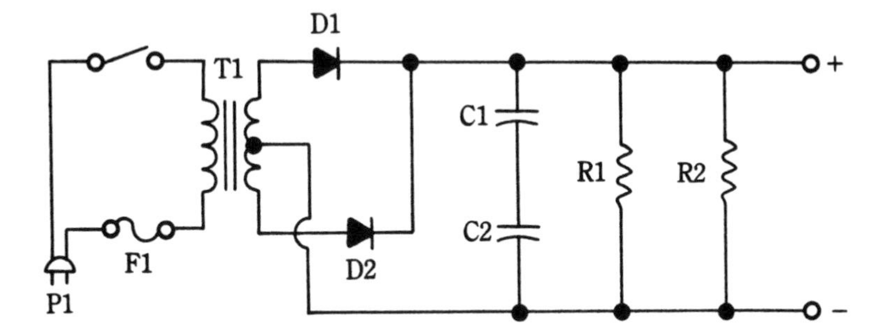

C1 - C2 - 100 Microfarad Electrolytic - 100 Vdc
D1 - D2 - 100 PIV 2 Ampere
F1 - 2 Ampere 250 Volts
P1 - Male Line Plug
R1, R2 - 100,000 Ohm, 1/2 Watt Carbon
S1 - SPST 3 Ampere
T1 - 115 Volt Primary - 20 Volt Secondary 1 Ampere

4-14 A dc power supply schematic and parts list.

each identical component, but the numbers advance in relationship to the total number of components used. Even though some components might be identical in value, they are still given separate numerical designations when combined in the components list as shown.

When building circuits from schematic diagrams, common sense must enter into the picture. It has been stated previously that single lines used in schematic drawings represent conductors. This is true in many instances. However, the conductor could actually be a part of a component lead. Whether or not a separate conductor is used to interconnect two components will be determined by how closely they are spaced during the construction process. I realize this sounds a bit confusing, but perhaps the schematic diagram in Fig. 4-15 will help clarify any misunderstandings.

The circuit shown includes three resistors, all of which are connected together in parallel. Taking the circuit quite literally, a conductor would connect the bottom of R1 to the bottom of R2. Another conductor would be used between the bottom of R2 and the bottom of R3. Two other conductors

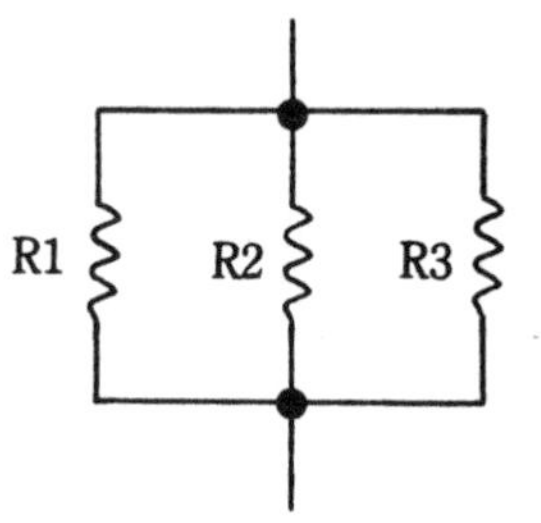

4-15 A simple circuit made up of three resistors connected in parallel.

would be used to connect the top leads of the components. In actual practice, this technique probably would not be done, since the three components would be mounted close to each other with the existing leads intertwined (Fig. 4-16).

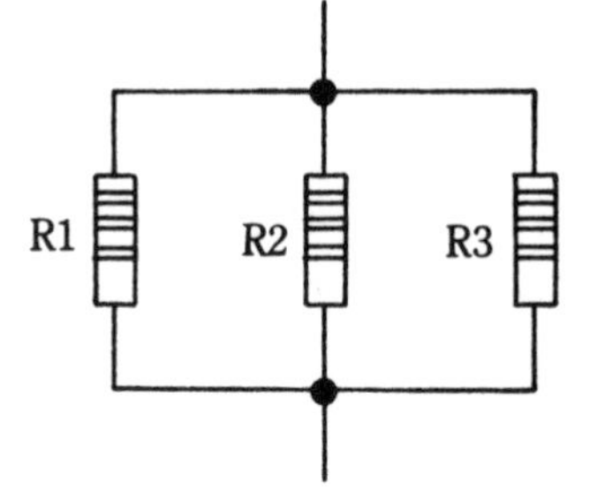

4-16 Pictorial diagram of the circuit of Fig. 4-15.

Naturally, you will want to make all electronic circuits as compact (and dependable) as possible by using a minimum amount of point-to-point wiring and trying to make the component leads serve for interconnection purposes. Of course, in the above example, if the three resistors had to be spread out over different parts of the circuit, then interconnecting conductors would be required. This type of building expertise comes with hours of practice in building electronic circuits of a simple nature and then moving on to more complex designs. Many excellent books out on electronic building are recommended reading, but the scope of this book is to teach schematic diagram reading and drawing, rather than building techniques.

Using schematics for troubleshooting

Although schematic diagrams are used initially to build electronic devices, they are an invaluable aid to troubleshooting equipment when problems develop. Knowing how to read schematic diagrams, however, is not enough. It is also necessary to know what electronic components do in a circuit and how various basic circuits operate. Remember, no matter how proficient you are at electronics troubleshooting, most repair jobs become real headaches without a good schematic representation of the equipment under test. Schematic diagrams clarify circuits. They present the various circuit elements in a highly logical and easy-to-understand manner.

When a circuit is built from a schematic drawing, it does not usually resemble the schematic physically. This was not true of the simple flashlight circuit discussed earlier, but will be true of most complex circuits. It is highly impractical to build a complex electronic circuit by placing the components in the exact positional relationship as is done in the schematic diagram. Schematic diagrams purposely spread out the components so the reason many schematic drawings are physically smaller than the finished device is because the schematic symbols are physically smaller than the components they represent. Naturally, schematic diagrams are two dimensional, whereas electronic components themselves are three dimensional. You need only to look inside the television receiver or other electronic device to realize the complexities that can be involved in troubleshooting without a schematic diagram to aid you.

If you know a bit about electronic components and how they operate in various circuits, then a schematic diagram can be used to indicate (without any equipment testing) where a particular problem might occur. Then, by testing various circuit parameters at these critical points and comparing your findings with what the schematic diagram indicates should be present, a quick assessment of the problem or problems might be obtained. For example, if a schematic diagram shows a direct connection between two components in a circuit and a check with an ohmmeter reveals a very high resistance between the two, then it can be assumed that a conductor is broken or a contact has been shaken loose. By the same token, if a schematic diagram shows a 100-ohm resistor between two components and the reading with the ohmmeter is very high, this might be an indication that the 100-ohm resistor has become defective.

Beginners to electronics troubleshooting and schematic diagram reading sometimes assume that a professional troubleshooter can immediately diagnose a particular problem (i.e., locate the bad part) by simply referring to the schematic diagram. This might sometimes be true of simple circuits, but is rarely so in complex designs. Often, the schematic diagram allows repairmen to make educated guesses as to where or what the trouble might be, but a true diagnosis usually requires testing.

The reason for testing is simple. A particular malfunction in an electronic device will not necessarily point to a single cause. Often there are many, many possible causes that the matter can be expanded from. For example, if a circuit will not activate and no voltage can be read at any contact point as indicated by the schematic, it can be safely diagnosed that no current is getting through the circuit elements. However, what has caused this failure of the current to flow? Has one of the components in the power supply become defective? For that matter, has the line cord been accidentally pulled from the wall outlet? Has a conductor broken between the output of the power supply and the input to the electronic device? Has the fuse blown?

Here the schematic diagram can be relied upon heavily, in conjunction with the various standard test procedures. The technician might wish to find the contact point that serves as the power supply output, indicated on the schematic drawing. If he tests voltage at this point and it appears normal, then he can assume that the problem lies at a point further on in the circuit. The schematic diagram and his test instrument readings allow him to methodically search out the problem by starting at a point in the circuit where operation is normal and proceeding forward until the point of inoperation is determined.

Using the same example, if there is no output from the power supply, the technician knows that he must search backward toward the trouble point. Chances are, he will continue testing until he reaches a point of normal operation and then proceeds forward from there. Using a schematic diagram, the technician might go all the way back to the original source of power, the 115-Vac household supply.

All of this could be done without a schematic diagram. However, in most circuits, the length of testing would be multiplied by many, many times. As you become more experienced in the art of electronics troubleshooting, the information contained in schematic drawings becomes more valuable.

Recall the flashlight circuit of Fig. 4-7. Although the schematic diagram does not indicate it, the two batteries in series should yield a dc potential of 3 volts. Some schematic diagrams do provide voltage test points and maximum/minimum readings. These will be discussed later.

Assume that the flashlight is not working and it is necessary to test the circuit with a voltohmmeter and this schematic diagram. First of all, we might measure the potential across the series-connected batteries. With the positive probe of the voltmeter placed at the positive battery terminal and the negative probe at the negative terminal, we should get a reading of 1.5 volts across each battery. If both read 0, then it can be assumed that both are discharged. If one reads normal and the other reads 0, then only one will have to be replaced. However, if both batteries read normal, then the next voltage probe might be accomplished at the light bulb itself. Here, a reading of 3 volts should be expected under normal operation. If you read 3 volts here, then you should immediately know what the problem is by looking at the schematic drawing. That's right! The bulb has blown. The circuit shows that the current path is through the light bulb. If current flows through this bulb, then it has to light. If voltage is available at the base of the light bulb, then current has to flow through the element . . . unless it has opened up.

On the other hand, suppose you get a normal reading at the batteries, but no reading whatsoever at the light bulb. Obviously, there must be a break in the circuit between these two circuit points. Three conductors are involved here: one between the negative terminal of the battery and one side of the bulb, another between the positive battery terminal and the switch, and still another between the switch and the other side of the bulb. Obviously, one of the conductors has broken (or a contact has been lost where the conductor attaches to the battery), or the switch is defective. Looking at the schematic, we can test for a defective switch by placing the negative voltmeter probe on the negative battery terminal and the positive probe on the input to the switch. If a reading is obtained here, then the switch is defective. If you still get no voltage reading, then one of the conductors is loose or has broken.

Admittedly, this is a very basic example of troubleshooting using a schematic diagram. But assume that the flashlight circuit is highly complex, one you know nothing about. Then the schematic diagram becomes an invaluable aid and a necessary adjunct to the standard test procedures with the voltohmmeter. This same basic test procedure will be used over and over again when testing highly complex electronic circuits. In most instances, no matter how difficult the circuit design appears to be, it is merely a combination of many simple circuits, and each will need to be tested individually.

Figure 4-17 shows a simple electronic circuit (actually, a portion of one) that has been presented in a form that will further aid the electronics troubleshooter. The circuit consists of a single transistor and a few other components. Note that test points (abbreviated tp) have been provided at several different locations.

When troubleshooting, the technician will apply his voltmeter between these points and ground and note the readings obtained. In many electronic circuits, actual voltages might

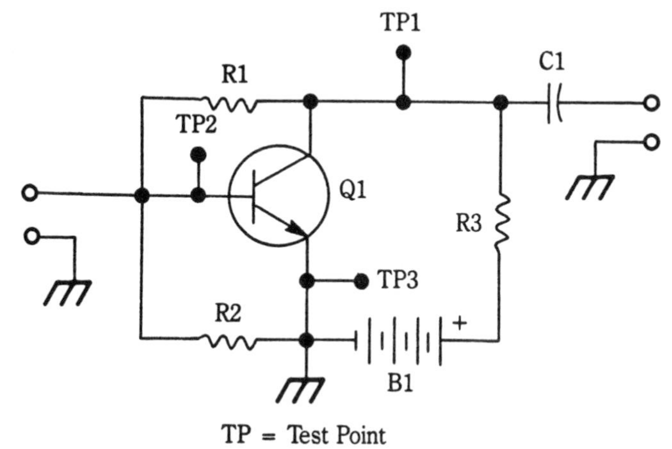

4-17 Simple electronic circuit diagram that also shows test point locations.

deviate from design values by 15 to 20 percent, but this information is usually contained at the bottom of the schematic drawing. If the readings obtained are within this known error range, then the technician can tentatively assume that this part of the circuit is operational. However, if the readings obtained are zero or well out of this range, then the technician can tentatively suspect a problem with this circuit portion or possibly other circuits that feed it. This method is another that symbolically illustrates electronic circuits and it does far more for the technician than a standard diagram.

Today many schematic drawings included with electronic equipment, especially the types that are built from a kit of parts, contain invaluable information that aids not only in troubleshooting, but also in the original testing and alignment procedures, which are often required when construction is completed. As a further aid, pictorial diagrams might also be included to provide even more help in the troubleshooting process. This subject will be dealt with in a later chapter.

Other forms of designation

Although it has been stated that it is standard practice to give every electronic component included in a schematic diagram its own alphabetic/numeric designation, a few other forms are also acceptable. Figure 4-18 shows a simple schematic diagram that eliminates the parts lists completely and contains no alphabetic/numeric designations. Here, the components are identified only by schematic symbols. However, value designations are written alongside each in many instances and in others, the actual component number is included. Using the example shown, we know that the transistor is a 2N2222 type and that the resistor has a value of 100 K ohms (100,000 ohms). Additionally, the capacitor is valued at .01 microfarad (μF). Sometimes, a statement will be contained at the bottom of the schematic diagram that includes information about these value designations. It might read "All capacitors are rated in microfarads (μF). All resistances are given in ohms."

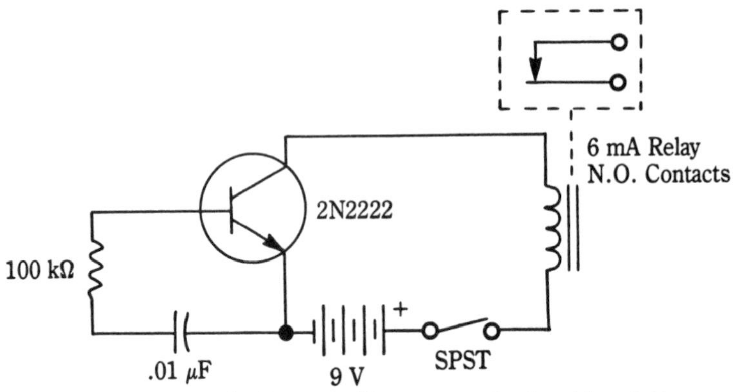

4-18 Simple schematic diagram with component values/descriptions, or in the case of the transistor, the actual component order number.

With this type of key, the person reading the schematic diagram will be dealing with numbers only. Given the same example, the resistor would carry a designation only of 100 K and the capacitor would be designated as .01. It is common practice in many schematics to use this type of designation routine. Some will even use the alphabetic/numeric system in combination with this latter system. Here a partial parts list is included, which is referenced by the alphabetic/numeric designations. All components containing only a numeric value designation will be excluded from this list.

Schematic/block diagram combinations

Sometimes a block diagram and a schematic diagram are combined. This method is shown in Fig. 4-19 and is used when a particular circuit is to be highlighted and explained, especially as to its relationship to other circuits. The figure shows a buffer amplifier that is used in a radio frequency transmitter. A full schematic diagram is provided for the buffer circuit alone, whereas block diagrams indicate its relative position in regard to the master oscillator and the amplifier.

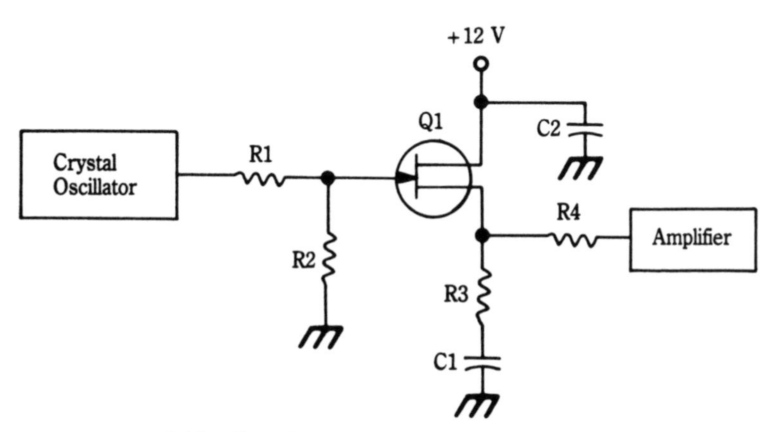

4-19 Combination block/schematic diagram.

This diagram serves two purposes. First of all, the person reading the schematic portion of the diagram can study the actual component makeup of the buffer circuit. He or she also is informed as to this specific circuit's place in the overall device. The block/schematic representation here indicates that the buffer receives its input from the master oscillator and channels its output to the amplifier. Another schematic diagram and block diagram combination might be used to describe another portion of this same device. Figure 4-20 shows an example whereby the master oscillator has been highlighted and indicates that the circuit channels its output to the buffer, which then inputs the amplifier. The only new

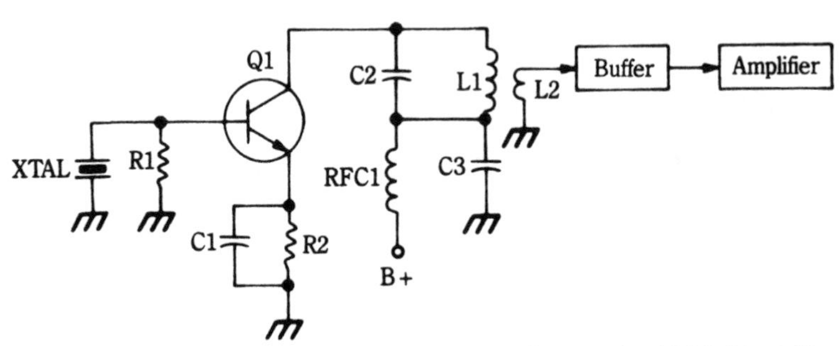

4-20 Another example of a block/schematic diagram that highlights a different portion.

information that is obtained here is contained in the schematic representation of the master oscillator.

Recall the basic block diagram of the ac-to-dc converter in chapter 1, Fig. 1-1. Figure 4-21 shows the schematic representation of that diagram. Comparing the two diagrams, note that in the schematic (Fig. 4-21), all of the actual components are shown rather than labeled blocks. Figure 4-22 shows how the block diagram (Fig. 1-1) relates to the schematic diagram. If you leave any stages out of the block diagram, you can be sure to catch this when you try to draw the schematic diagram.

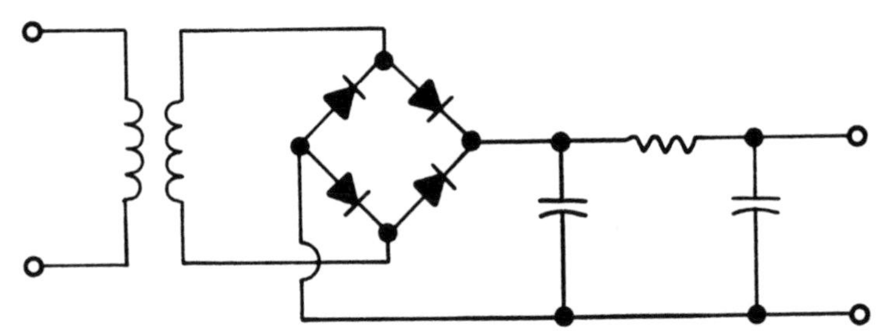

4-21 Basic ac-to-dc converter schematic from the block diagram of Fig. 2-1.

To briefly explain the circuit, the input is still at the left. The signal goes to the transformer to set up the proper ratio for conversion to a dc signal. The four diodes in the diamond configuration constitute the ac-to-dc rectifier, which does the

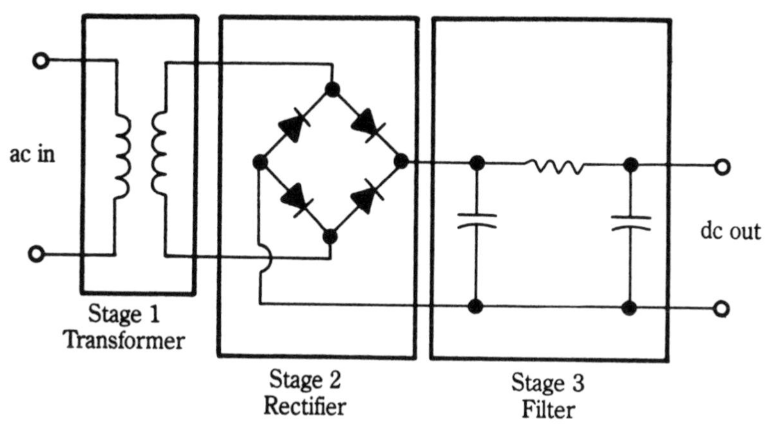

4-22 How the block diagram of Fig. 2-1 relates to schematic of Fig. 4-21.

actual conversion. The final stage, the ripple filter, acts to smooth out the signal after the conversion before it reaches the output at the right of the circuit. This circuit is referred to in pictorial form in chapter 6.

Summary

This chapter has provided a brief sampling of simple electronic circuits, showing how they can be read and/or drawn by viewing the actual completed circuit. After a bit more training, you should be able to view a simple schematic drawing and be able to visualize what the finished circuit will look like.

Using the road map as an example again, we all know that a roadway does not actually appear like the black line which is used as its symbol. We can actually visualize a secondary road or even a superhighway and do so whenever we see their symbols. The same is true when reading schematic diagrams. When a capacitor symbol is indicated, we can visualize the actual component and maybe even make mental notes on the physical aspects of construction. Symbology becomes a second language, one we can begin to think in. When this level of proficiency occurs, no great mental distinctions will be made between a schematic diagram, a pictorial circuit drawing, and the actual completed circuit.

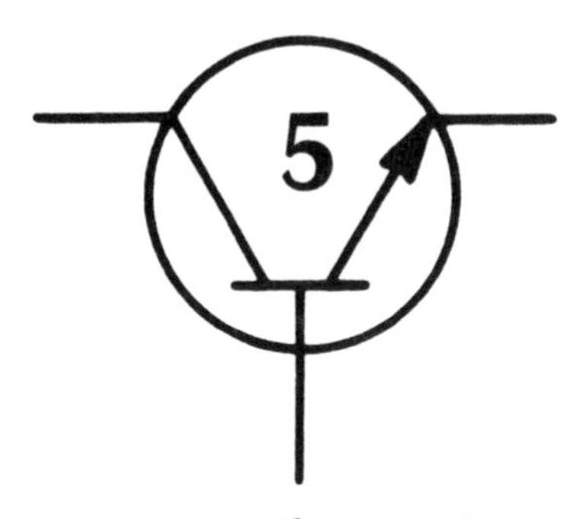

Combining simple circuits

WHEN LEARNING TO READ AND WRITE SCHEMATIC diagrams, some people are often misled by the apparent complexities involved. "Sure, anyone can learn to read schematic diagrams of simple circuits, those which contain a transistor or two, but it must be nearly impossible to learn to decipher more complex schematic drawings in a short period of time." This statement is usually not true. There is really no such thing as a highly complex circuit.

Circuits that appear to be very complex are always made from building blocks, and each of these is a simple circuit. A device that contains twenty transistors and associated circuitry might really be twenty simple circuits, each using a single transistor, that have been combined. Certainly, at first glance, these scientific-looking scratchings might appear undecipherable, but if you just look a little while longer, you will begin to identify the simple circuit components. You will even begin to see how the simple circuits are combined.

To illustrate this point, Fig. 5-1 shows a simple circuit for a crystal radio constructed with an antenna, a coil, a variable capacitor, and a diode. Certainly, this illustration could be classified as a simple schematic drawing because the circuit is so simple that it only contains three components in addition to an antenna. If you add another component to this circuit, you

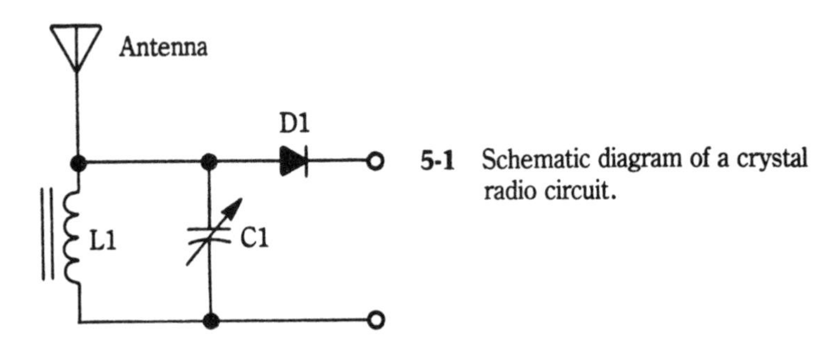

5-1 Schematic diagram of a crystal radio circuit.

can actually receive nearby AM radio broadcasts. This fourth component is a crystal headphone. However, the purpose here is not to build a simple crystal receiver, but to build an AM radio detector whose output is amplified.

Figure 5-2 shows another simple schematic diagram, a one-transistor audio preamplifier circuit. It consists of only six components: two capacitors, two resistors, a transistor, and a 9-volt battery. This circuit has the ability to accept low-level audio signals as input and to pass the same equivalent signal at a higher level as its output. This circuit, again, is a very simple one that can be drawn in a couple of minutes and probably built in ten minutes or so.

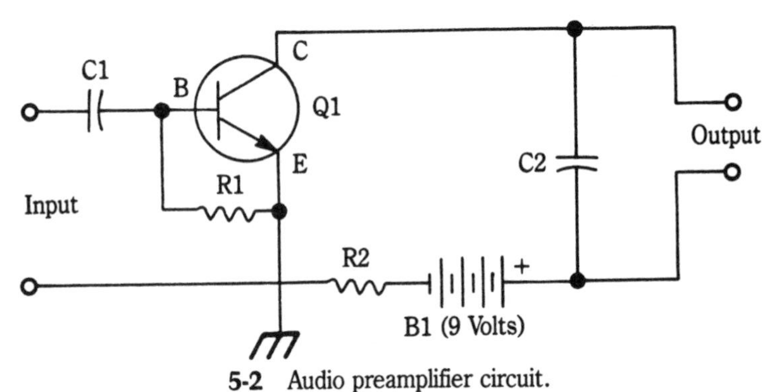

5-2 Audio preamplifier circuit.

Figure 5-3 shows a more complex circuit. But is it really? This circuit is composed of the previous two simple circuits, which have been combined to form an amplified crystal radio or a detector with a preamplifier, as it might be known. The dotted line indicates where the two circuits have been com-

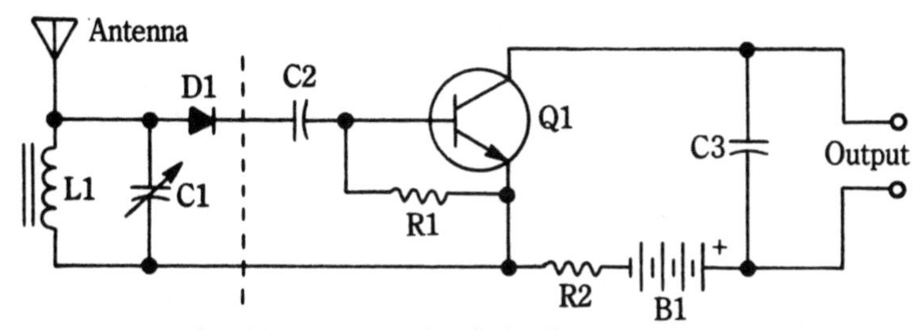

5-3 Combination crystal radio/audio preamplifier circuit.

bined. If you were able to decipher the symbols in the two previous schematic drawings, you can certainly decipher this one.

The last circuit shown still has a relatively low audio output level, although it is much higher than that which would have been obtained without the transistor amplifier stage. The level would not be sufficient to allow connection of a loudspeaker, so the circuit in Fig. 5-4 can be used to increase the output signal to a point where the audio can be heard in a speaker. This circuit is a basic example of the stages that are present in your AM table radio.

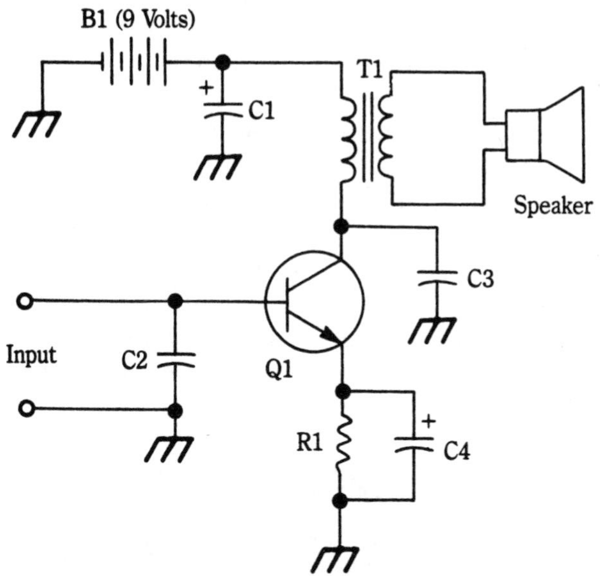

5-4 Audio amplifier circuit.

This last circuit is an audio amplifier and will accept the output from the one-transistor preamplifier and increase the signal even more. The output from the original crystal radio would not be sufficient to drive this latter circuit, so the preamplifier stage was absolutely necessary. Notice that this audio amplifier circuit also requires a 9-volt power supply; it can get its power from the 9-volt battery used for the preamplifier. Figure 5-5 shows the completed circuit, which looks even more complex, but really is not. It is merely a combination of three simple electronic circuits, all accomplishing different functions.

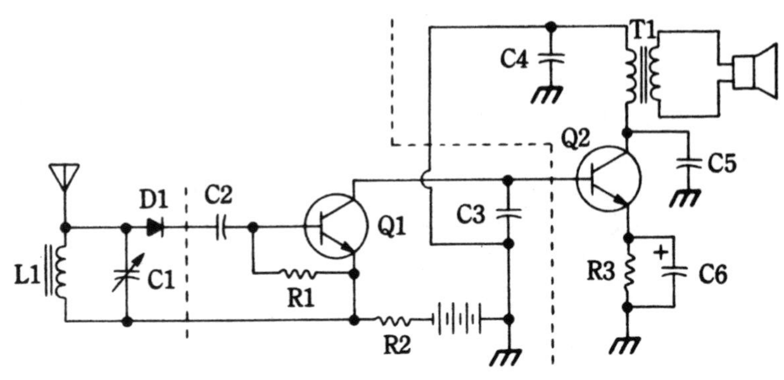

5-5 Complete radio circuit.

The basic process used to make all electronic circuits is a process of combinations. First, electronic components are *combined* to form simple circuits. Then, simple circuits are *combined* to make complex circuits. Complex circuits are *combined* to make multi-purpose/multi-functioning electronic equipment and devices. Even these devices can be *combined* to form systems. Systems are *combined* to make complex networks. This step is as high as I'll take the combining process, but it can even go considerably beyond this point.

Analyzing larger schematics

Figure 5-6 shows a rather complex-looking circuit of an AM transmitter. However, it has been blocked off into six different sections, each of which is composed of a simple electronic

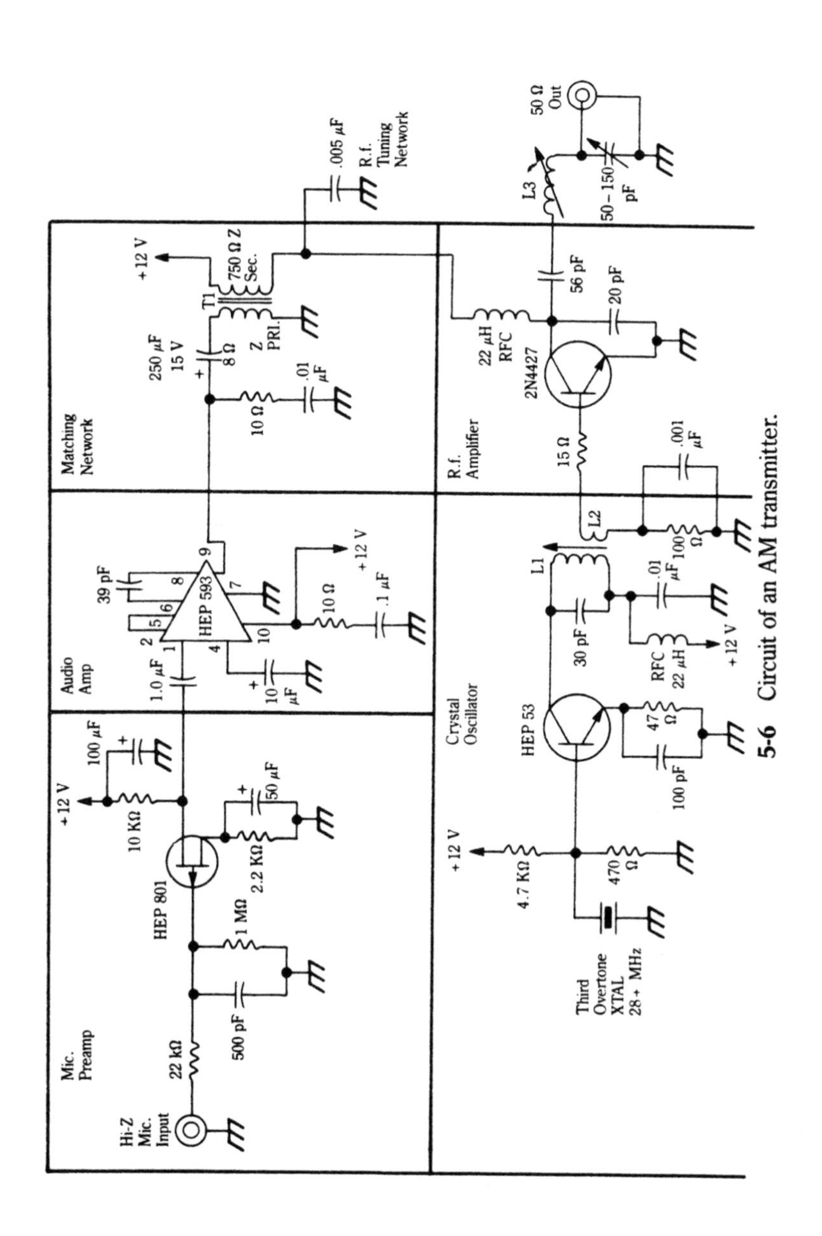

5-6 Circuit of an AM transmitter.

circuit. It might be necessary for the beginner to block-off sections of circuits such as this one in order to gain a better understanding of what is taking place.

Examine this circuit closely and see what is occurring. The first section (top left) is the *microphone preamplifier*. This simple circuit consists of a single p-channel field-effect transistor, along with a few resistors and capacitors. The purpose of this circuit is to increase the level of the signal which is input to the circuit by the microphone.

The next section is the *audio amplifier*, which is made from an integrated circuit and a few extra capacitors and resistors. The audio amplifier increases the amplitude of the output from the preamplifier, just as was the case in the simple receiver. At the output of the audio amplifier, there is a matching network consisting of T1, a modulation transformer.

Now, moving to the bottom left of the drawing, is the radio frequency portion of the circuit. The previous portions all dealt with audio frequency. First is the *crystal oscillator circuit*, which establishes the output radio frequency. This section again is a simple one-transistor circuit that is coupled to a master network.

At the output of L2 is another amplifier stage. The *crystal oscillator* increases the signal established by the third overtone crystal, which is at radio frequency. The microphone preamplifier causes the same effect, but at audio frequency. The *rf amplifier* boosts the signal output from the crystal oscillator, just as the audio amplifier did in the microphone circuit. Finally, the *tuning network* at the output of the rf amplifier matches this output to the antenna. This is what the matching network did at the output of the audio amplifier, but again, at audio frequencies.

Look at the overall circuit again. It's not all that complex, is it? Of the six stages, only four contain active components (transistors and integrated circuits). This apparently complex circuit is merely a combination of six different simple circuits, any of which can be quickly read and easily understood. Your proficiency at this point might not enable you to explain each circuit portion; however, you should be able to identify all of the schematic symbols in this figure, even if you have to refer to the schematic symbols table in Appendix A.

Knowing the symbols is extremely important, and after a while, you will begin to recognize combinations of symbols as being a specific type of circuit (amplifier, oscillator, matching network, etc.). For example, the crystal oscillator circuit is identified by the fact that the schematic diagram shows a third overtone crystal (XTAL) at the base of the HEP 53 transistor.

Generally speaking, circuits will start on the left-hand side of the page and progress to the right. This rule is not always true, especially when schematics are used to describe highly intricate devices, but is often so. This schematic follows the left to right routine. Although six basic circuit sections are in this diagram, there are two complex sections. The audio portion of the circuit (on the top) is one of them and the radio frequency portion on the bottom is the other. Notice that the audio section starts on the left with the microphone input and progresses through to the right with the audio amplifier and matching network. At this point, the audio section of this circuit is complete.

The radio frequency section reverts back to the left and moves forward to the 50-ohm antenna output on the far right. You could encounter the same circuit handled in reverse order (right to left). Here, the entire schematic is simply flipped over (end over end), so the input portion would start on the right and move to the left. This method usually won't be encountered.

Schematic diagrams follow logical circuit order. Using this same diagram as an example, you can see that there is no direct wiring connection between the microphone preamp-amplifier and the matching network. Just an audio amplifier is in between. You could put the audio amplifier section in the third top block and move the matching network back to the second block. However, this action would necessitate extending the wiring from the first block to the third block, temporarily bypassing the second block, which would then be connected to the output of the audio amplifier in the third block. This design requires highly complex drawings and diminishes intelligibility. A schematic diagram is supposed to clear up misunderstandings, not add to them. The exact same comparison can be made to the rf section below.

The schematic diagram usually follows the logical amplification or processing order of the input signal. This simple transmitter receives its audio-frequency (voice) input from a microphone. This signal is immediately amplified by the microphone preamplifier circuit, so it fills the first block. Next, the preamplifier-boosted signal must be further increased by the audio amplifier circuit, so this one is next in line. After final amplification, the output impedance of the audio amplifier must be matched, so the matching network fills the third block. You see, it's all logical if you take the time to understand the schematic drawing process and how it relates to the circuit it so efficiently describes.

Figure 5-7 shows an antenna matching circuit known as an *L-network*. It derives its name from its schematic appearance, which resembles the letter L. Another matching circuit, shown in Fig. 5-8, is called a *pi-network* because its schematic appearance resembles the Greek symbol for pi (π). Both of

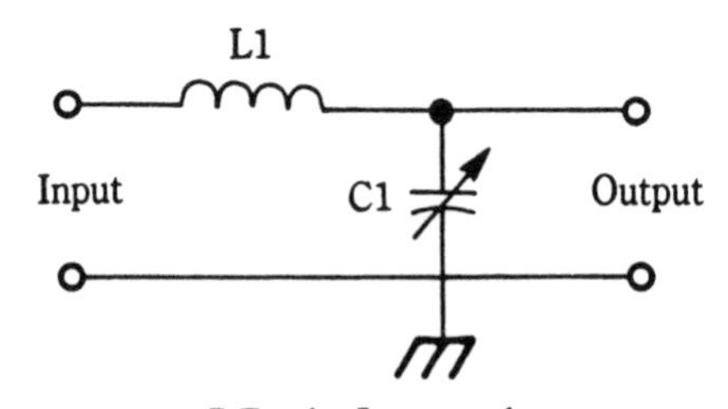

5-7 An L network.

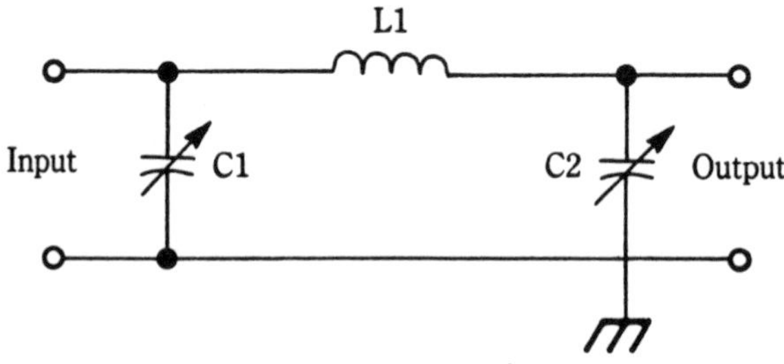

5-8 A pi network.

these are very simple circuits that can be combined to form a ***pi/L-network*** as shown in Fig. 5-9. Here, the output from the pi-network is simply fed to the input of the L-network. This latter circuit seems more complex than either of the two, but it is simply a combination of them.

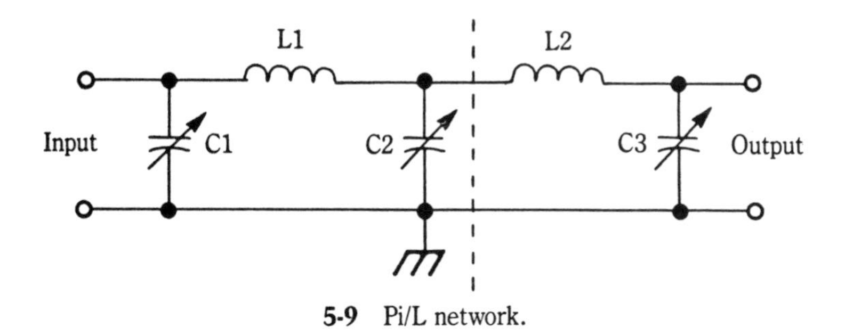

5-9 Pi/L network.

Figure 5-10 shows a code-practice oscillator circuit designed to operate from a 9-volt battery. Notice that the key symbol is used to indicate the code key, and that there is a single transistor whose circuit is composed of a few additional components, then the 9-volt battery. However, suppose it is desirable to replace the 9-volt battery with a 9-volt dc power supply that can be built from the circuit shown in Fig. 5-11.

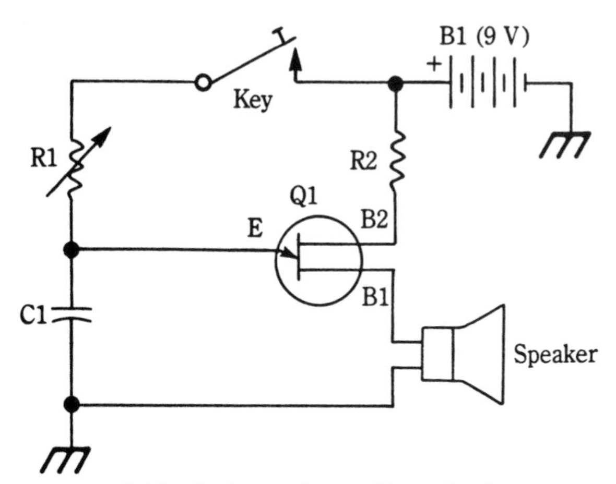

5-10 Code practice oscillator circuit.

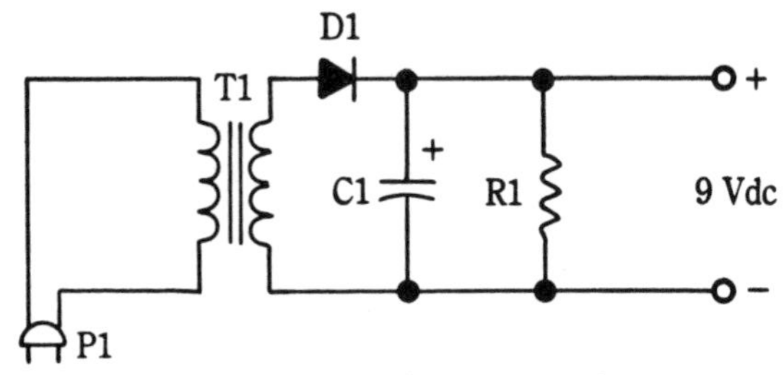

5-11 Nine-volt dc power supply.

Here, we have two discrete circuits: one is an oscillator and the other is a power supply. Both can function independently, although the oscillator will require a 9-volt battery in this case. However, Fig. 5-12 shows how the two have been combined to apparently form one complex circuit. This same schematic could also be presented as shown in Fig. 5-13. Here the two circuits are included on the same page, but are

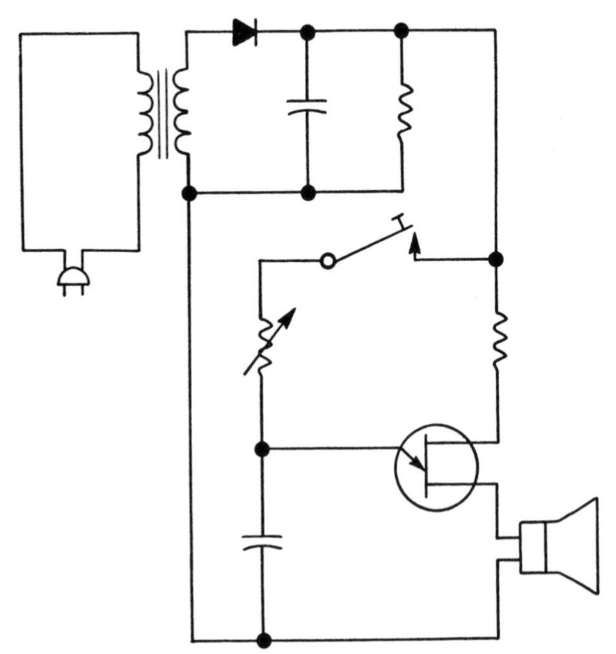

5-12 Combination of the code practice oscillator and the dc power supply circuit.

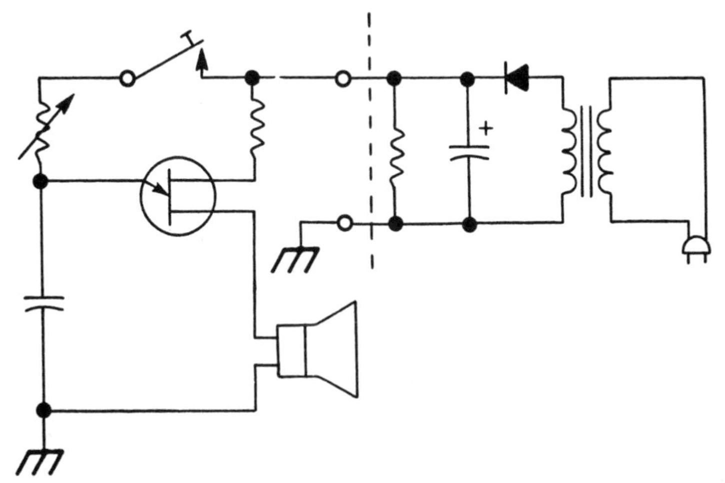

5-13 Alternate method of representing the circuit of Fig. 5-12.

kept separate by spacing. A cable with matching connectors is used to allow the power supply to provide current to the oscillator. This latter circuit might be a little less confusing, but it is electrically equivalent to the previous one.

Figure 5-14 shows a filter circuit that has an input and an output. The inductor and capacitor (in practice) will be chosen to present a certain value that would allow some signals to pass and others to be blocked. Since this is only a simple single circuit, it can only be set up to handle a certain range of frequencies based upon the component values of the inductor and capacitor.

Figure 5-15 shows a complex circuit composed of many filters, each of which can have a different capacitance/inductance value in order to respond to different frequencies.

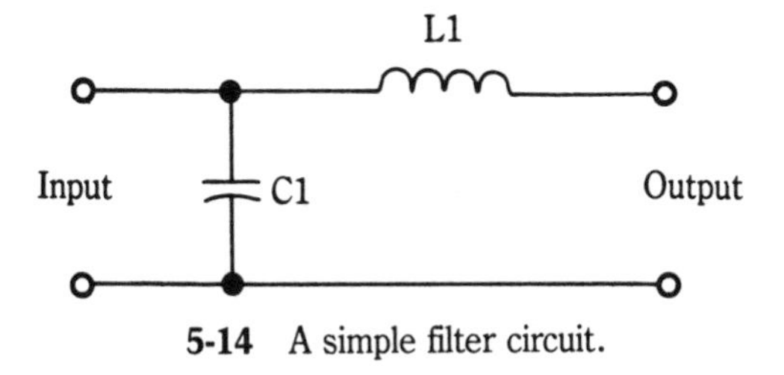

5-14 A simple filter circuit.

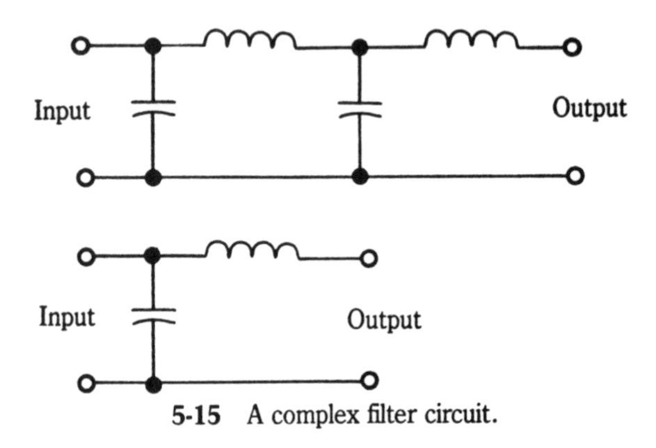

5-15 A complex filter circuit.

Naturally, the circuit looks more complex, but in reality, it is just a repetition of the basic circuit in Fig. 5-14. However, different component values can be chosen for each inductor/capacitor combination. You will find this repetition in many electronic circuits and their equivalent schematics. Sometimes it is necessary to have three or four of the same circuits arranged in parallel or series to bring about a particular electronic result.

If you know how one circuit operates, then you know the basic operation of them all. A problem that occurs in one circuit might also occur in all the others, and the schematic diagram can be used for tracing purposes. For example, if through testing you know that a particular oscillator changed frequency because of a defective grid resistor, then should another oscillator exhibit the same characteristics of improper operation, the schematic can be consulted, the grid resistor located, and a further test made. Without the schematic drawing, it would be very difficult to find the grid resistor quickly.

It can be further stated that this single grid resistor, which might cost ten cents or so, could be causing a highly complex system to fail. By isolating a complex system of circuits into a number of complex circuits and then breaking down the complex circuits into simple circuits, schematically-aided troubleshooting becomes less difficult.

Breaking a system down into complex circuits, makes it easier to determine which circuit might be creating the problem for the entire system. This complex circuit is then broken down into simple circuits, and a determination is made as to which simple circuit might be at fault. The simple circuit is then broken down into separate components, which are then examined individually for a possible fault. Through this systematic method of elimination (all done schematically), the equipment is repaired.

Many failures of highly complex pieces of electronic equipment are from a problem with a single component. Sometimes this failure will cause other components to fail as well, but the repair must start with the first failure. Occasionally, two simultaneously defective components will be discovered, but this is a rarity. Certainly, it is necessary to become familiar with the equipment you are attempting to repair, but after normal operation is studied and understood, a schematic drawing can often be used to identify the general area of any future faults. From this method, highly suspect discrete components can be identified. It is then a matter of getting into the equipment with a meter or other test instrument to check these suspicious components. Even with the finest test instruments, it is almost impossible to quickly identify defective sections and especially components without a schematic diagram because you simply don't know where to look.

Getting comfortable with complex schematics

No matter how avid a hobbyist you might be, it is not possible for a beginner to sit down and simply read complex schematic drawings. You've got to start at the beginning. First of all, you must make certain that you know *all* of the schematic symbols you will be dealing with. Complex schematics are good for this purpose because they often contain a myriad of symbols, some of which might be unknown to you. Although you can't sit down with a complex schematic at the start and understand everything that's going on, you can use one or more to aid you in learning symbology.

Once you feel comfortable with the schematic symbols, put away the complex schematics and start poring over the simple ones. Many of the simple projects books that TAB Books offers are an excellent source. However, you will also find them in magazines geared to the electronic enthusiast, especially the beginner. Don't study just one type of schematic (those which deal with oscillators, for example). Check into oscillators, amplifiers, solid-state switches, tube-type circuits, rf circuits, audio-frequency circuits, etc. You will find that many of them are basically alike, with only a few changes in component values. When you can identify an amplifier circuit by looking at the schematic alone, then you are making progress.

It is impossible to learn to read and draw schematic diagrams without learning a great deal of practical and theoretical electronic design and application. Too many people feel that all one has to do to become a crack theorist or troubleshooting technician is to learn to read schematic drawings. This statement is not true. Just knowing how to read a road map doesn't qualify you to drive an automobile from coast to coast. It is necessary also to learn how to drive a car; this allegory parallels with the need to learn to draw schematics. The schematic diagram is an aid to the understanding of electronics and electronic circuits. Like the road map, throughout your electronic pursuits, the schematic diagram will be a constant aid, giving you an indication of what could be wrong and where the problem is found in the circuit.

Once you can comfortably identify simple electronic circuits from schematic diagrams, it is time to move on to more complex drawings—not *too* complex, however. If you try to move too quickly, you might become frustrated and give up altogether. Your next step will be an intermediate one and will include the circuits that combine a few of the simple circuits you have previously studied. Sometimes additional components are added to match the output of one circuit to the input of another.

Now it is time for a new phase of your course in reading schematic diagrams. Try to select books and publications that offer a simple, theoretical, and practical discussion of the circuit that the schematic depicts. Again, electronic projects books are ideal because they usually include a schematic diagram along with an explanation of the basic circuit functions.

Even better, begin to build some simple circuits in a workshop at home. Many books have been written about building electronic circuits. Most projects books include the basics in the front and the circuits in the back portion, so these are quite handy. You will be surprised at how your first electronic circuit looks when compared with the schematic drawing. Your study will continue from this point by examining the functioning circuit and noting the relationship of the physical components to those in the schematic drawing. You can further your electronic knowledge by experimenting with these circuits, substituting different components, for example. You might even be able to improve the circuit operation. All improvements should be duly noted and a new schematic can be drawn up indicating your changes. You might want to simply pencil in the changes on the schematic diagram you were building from.

Now, when you feel comfortable building electronic circuits from simple schematic diagrams, you might want to combine two or more into a single circuit. Take two schematic diagrams from a projects book and combine them on paper. You will have to draw your own schematic to serve as your plan for the following building procedure. You might even be knowledgable enough this time to provide some additional circuits that might be needed to interface the two. Combining electronics building with learning to read schematic diagrams is the most efficient way to improve your electronics knowledge. Admittedly, it can be quite boring to pore over schematic diagrams for hours on end. However, when you can refer to a portion of a schematic drawing and then wire the components in place, much of the boring aspect is removed and knowledge is more efficiently retained.

Before you know it, you will have obtained a great basic knowledge regarding the schematic diagrams and the building of electronic circuits. The circuits that you used to think of as being complex will soon seem like old friends. *Caution:* at this point in your development, you might have a tendency to stick only with the circuits you know best. Don't let this happen. As soon as you reach one stage of comfort, move on to diagrams that are more difficult and make you feel uncomfortable again. If you don't do this, you will be stuck at this one stage of development for a long time. Although it might not be necessary, try to continue building the more complex projects. Admittedly, this practice can become expensive, so if building is not possible, continue to read schematics and decipher the various circuit components anyway.

You will continually be surprised at what you know and what you don't know. For instance, many people fairly new to electronics feel that a commercial AM radio transmitter must be a highly complex device. Most are surprised to learn that it is technically less complex than the transistor pocket receiver you use to detect the broadcasts. By comparison, a commercial transmitter is a simple circuit, even though it might be seven feet high and just as wide. This transmitter size is directly related to component size. For example, a modulation transformer in a commercial transmitter might weigh several hundred pounds. This and other components make this equipment necessarily large. However, a modulation transformer for a walkie-talkie may weigh less than an ounce. Schematically, both modulation transformers will be drawn in an identical manner. Circuit complexity, then, is unrelated to the physical size of a component. Complexity is based on the number of components as well as the number of different combinatorial circuits that are included.

A schematic diagram presented earlier in this chapter showed the circuit of a simple radio transmitter. The circuit of a commercial radio transmitter would not look much different, although the components specified might be a hundred times the size of the ones used for the previous circuit.

It's not usually the massive pieces of equipment that are the most complex, both electronically and schematically. The tiny units that can be held in the palm of the hand often take the prize for schematic complexity. For this reason, the beginner to electronics and schematics should not shy away from any particular circuit, device, or equipment just because there is an impression of tremendous complexity. You might be wrong, but even if you aren't, every schematic diagram will contain portions that you will be able to comprehend. After you have passed the intermediate stage of learning schematics, then the complex circuits are next. The first thing to do is to break them down into intermediate circuit sections and again into simple circuits. Try to obtain schematics of a complex nature that also include a thorough explanation of how the circuit functions.

Recall the block diagram of Fig. 2-2, the strobe light circuit. Compare it to the schematic in Fig. 5-16, which has all of the components depicted. The circuit is powered with 120 Vac, which comes in at the left side of the schematic. The three terminals of the 120 Vac line go three separate paths via the color-coded wires: part of the signal goes to the fuse, part of it goes to the timing components, and part of it must go to ground.

Following the top signal path, power passes through the fuse only if the switch is closed. It then passes through the switch and the rectifier, after which part of the signal goes to the A terminal of the strobe light and part goes to the adjustable timing components. The adjustable timing components determine the frequency at which the light flashes (i.e., how long it stays on and off). This place in the circuit also serves as a junction where the bottom portion of the circuit interacts with the top portion to provide the proper signal to the remaining two strobe lamp terminals. This schematic is pictorially illustrated in chapter 6.

Finally, Fig. 5-17 shows the schematic for the power supply whose block diagram is shown in Fig. 2-3. Even though the schematic might appear to contain a lot of individual components, you should be able to determine which components

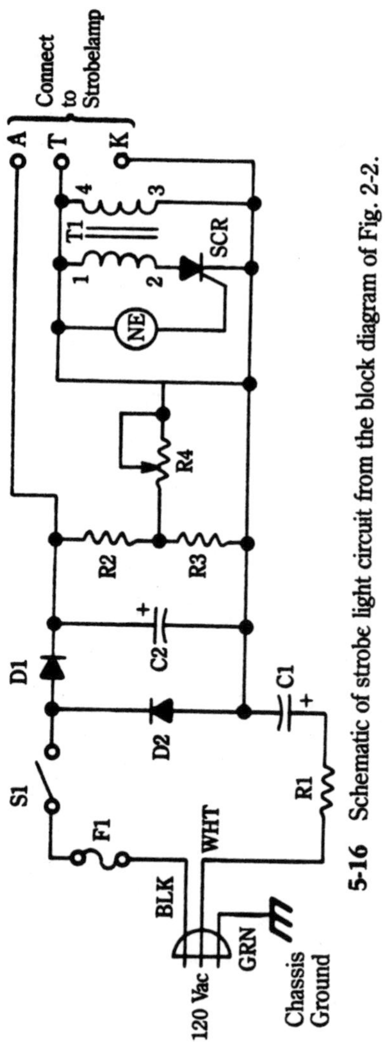

5-16 Schematic of strobe light circuit from the block diagram of Fig. 2-2.

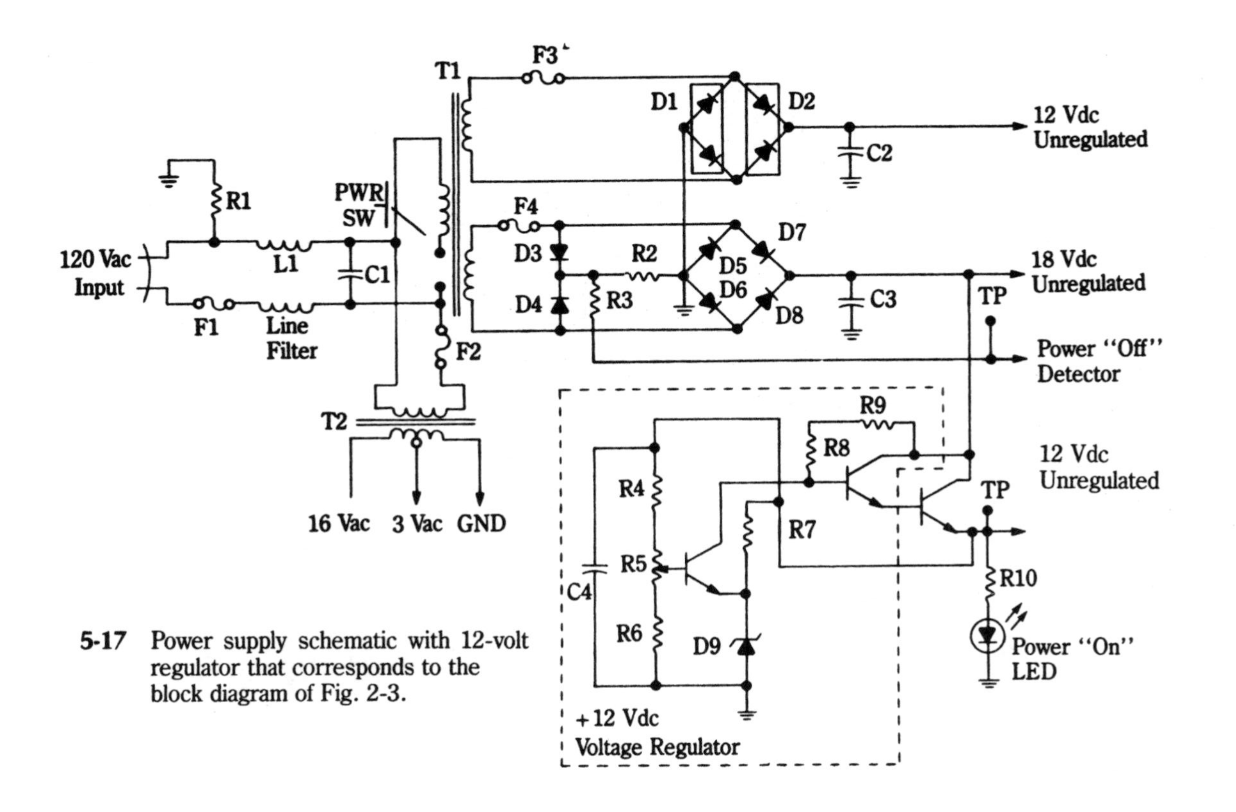

5-17 Power supply schematic with 12-volt regulator that corresponds to the block diagram of Fig. 2-3.

compose which blocks. The signal path begins at left, as usual. It passes through the line filter, which consists of the two inductors and the capacitor, C1. The signal then goes to two transformers. The bottom transformer provides two different ac voltages and a ground reference. However, the other transformer is followed by two rectifiers, as shown in the block diagram; in this case, the rectifiers are bridge types made up of four diodes apiece. Each bridge rectifier provides a different rectified (dc) voltage, both of which are different from that of the other transformer.

Part of the signal is tapped off of the top transformer (before it becomes rectified) to act as an indicator that denotes a "power-off" situation to the remainder of the device. For example, when this voltage reaches a certain level, it could be applied to a sensing circuit that would stop the machine to prevent it from being left in the operating mode in the event of a power loss.

The signal from the second rectifier, 18 Vdc regulated, also goes to the voltage regulator to provide a +12 Vdc regulated output. Therefore, this power supply generates seven different outputs that can be routed to various other circuits or systems to power a complex piece of equipment. The signal from the voltage regulator also turns on a "power-on" LED when the power supply is activated.

Summary

Reading and drawing schematic diagrams involves the methodology of breaking down complex circuits into simple ones. This is the way even the seasoned professional goes about this procedure. The beginner must do the same thing, looking at the part, rather than the whole. As a complex schematic drawing is systematically studied, the relationship of one circuit to all the others will eventually become apparent.

When using schematics for electronics troubleshooting, it is often unnecessary to understand the function of all circuits; only the ones that are potential trouble spots are important. Learning to read and write schematic diagrams is very similar to learning to receive and send Morse Code. Morse Code is a language of audible symbols, just like schematic drawing is a language of printed symbols. Once you learn either language, you are capable of communicating in it and allowing it to communicate with you.

Using Morse Code as a further example, a five-minute sequence of dots and dashes will mean nothing unless you are capable (initially) of breaking the data down into individual code words. As proficiency increases, the person deciphering this code stops hearing mere dots and dashes and begins hearing letters instead. It's at this point that you begin to think in the new language. As proficiency continues to increase, entire words are heard instead of the letters that form those words. Eventually, one will hear entire sentences and finally, it's just like a second language.

Reading and writing schematic diagrams follows a similar pattern of development. At first you will see individual components. Later, you will begin to see complete simple circuits hidden within complex circuits. Then, complex circuits can be identified and deciphered at a glance. Finally, you will begin to envision entire systems. This knowledge will not come quickly, but your proficiency will improve every time you practice if you push yourself into areas that are continually unfamiliar to you. This is a step-by-step process, one that proceeds as fast as the student is able to absorb the information presented.

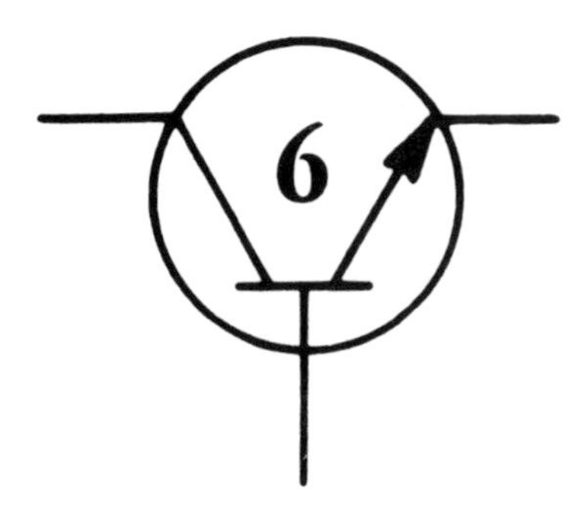

Pictorial diagrams

THROUGHOUT THIS BOOK, IT HAS BEEN STRESSED THAT symbology is the key to understanding how to read and draw schematic diagrams. Symbology is ideal for expressing the complex working of electronic circuits in a medium that can be readily accepted and processed by the human brain. Sometimes, however, we can use another type of diagram to aid in understanding circuits. At several points in this book, pictorial diagrams have been used to show the components that our schematic symbols represent.

Pictorial diagrams are often used in conjunction with schematic drawings to indicate physical relationships, which schematic diagrams don't really do. Rather, schematics show electrical and electronic relationships. Theoretically, an electronic circuit that can be built on a piece of circuit board one-inch square could also be built on a circuit board one-mile square. The components would simply be spread out and great lengths of hookup wiring would be used to interconnect them. From a practical standpoint, the resistance losses that would be incurred here would not allow the circuit to operate, but from a theoretical standpoint, physical relationships (component placement) are not usually dealt with.

When building or servicing some electronic circuits, the physical relationship of one component to another plays a very important role. Conductors will have certain inductive and capacitive effects and will act directly upon the circuits. The actual effect will depend upon the frequency of operation and many other factors. If two components are to be positioned very close to each other, the schematic cannot show this in a practical manner. Certainly, an English language notation can be made on the schematic drawing, but the more efficient method uses a pictorial diagram to show the actual components in the circuit.

Types of pictorials

Pictorial diagrams can be two dimensional in appearance or three dimensional. Some are drawn to look exactly like the finished circuit. Others are drawn in a quasi-block diagram form that roughly indicates the physical size of components along with physical spacing.

Figure 6-1 shows a schematic diagram. This diagram provides all of the electronics relationships needed to build a working model, but tells the builder nothing about how the components are to be physically situated. This particular circuit is best built on a small piece of perforated circuit board, so the drawing shown in Fig. 6-2 was also included to allow the builder to see how this latter job is to be accomplished.

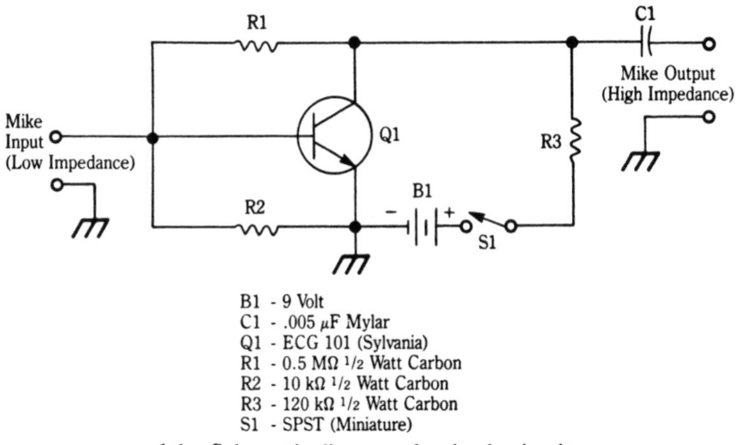

6-1 Schematic diagram of a simple circuit.

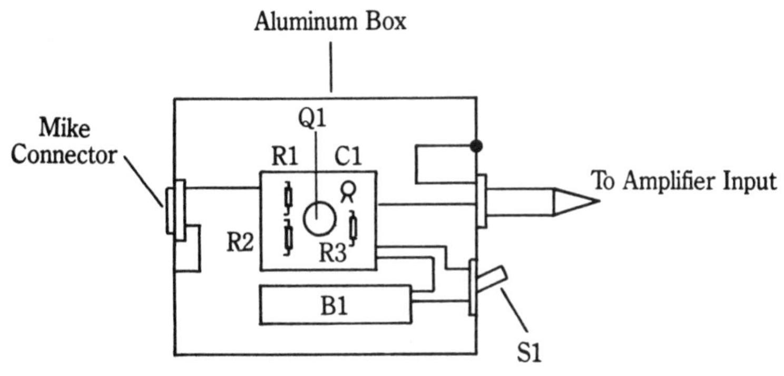

6-2 Pictorial diagram of the circuit of Fig. 6-1.

Notice that the component lists are also included so that you can cross-reference between the two diagrams.

This illustration is a pictorial diagram and it can be classified as being basically two-dimensional. Height and width are indicated in the drawing, but depth is not. The circuit board is in the exact center of a box that represents the aluminum compartment the overall circuit is to be mounted in. The physical placement of the mike connector, the switch, the circuit board, the battery, and the output plug are all indicated. This one drawing allows the builder to see how the five components should be situated on the circuit board and how the other components could be mounted to the box. Now, the builder really has something to work from—a schematic diagram *and* a simple pictorial drawing.

Although the preceding example is a pictorial drawing, the components shown do not look exactly like they really will. This language is still one of symbology, but the symbols are much more closely related to the general physical appearance of each component. The builder now has an idea of what his circuit should look like once it is completed.

Electronic components are usually very simple to draw in this manner. A pictorial of this type shows an overhead view, which is necessary when only two dimensions are to be indicated. A side view of this circuit would require that height (or length), width and depth be displayed.

Recapping just a bit, a schematic diagram gives the electrical and/or electronic relationship of the components in the circuit, and the pictorial drawing shows the physical relationship. Look at the last figure again. It shows little of the electronic relationship between the components in the circuit. The only electronic relationship that can be shown involves the battery (B1) and switch (S1). We can see that one lead of the battery is connected to one switch contact. The remaining battery lead and switch contact lead enter the circuit board at this point. Looking back at the schematic diagram of the same circuit, you can see that the positive battery lead does indeed connect to one of the switch terminals. The remaining switch terminal is connected to resistor R3. In the pictorial drawing, however, R3 is part of the circuit board and the physical relationship stops pictorially before this component is encountered.

If you had experience at building simple electronic projects, then the pictorial drawing would not be necessary in this case. Those who do not have many hours of bench experience, however, probably could not build the circuit with the schematic diagram alone. From a service standpoint, suppose this circuit was built and then failed suddenly. Assume also that a single resistor has burned up as a result of an overload and its value could not be identified. By referring to the pictorial diagram, you could quickly locate the burned up component and then reference it to the schematic diagram, which would in turn reference it to the components list. A new resistor could then be purchased and wired back into the circuit using the two types of diagrams.

How do you draw pictorial diagrams? Obviously, it is necessary to have a finished circuit on hand or to be experienced enough to accurately visualize it in your mind. Assuming that you have a circuit on hand, all you do is visually examine it and then start building your pictorial diagram, one component at a time. This method is the same one used to read and draw schematic diagrams. If the entire circuit is to be mounted on a piece of circuit board, then start by drawing a likeness of the board on paper. If the original board is ten inches by five inches, it might be necessary to reduce it somewhat. A representation that is five inches by two and a half inches will be exactly half the size of the original board. Therefore, all electronic components should be drawn half-scale as well. In practice, the exact dimensions are relatively unimportant. Simple circuits don't usually require strict adherence to physical placement, which is accurate down to a fraction of an inch.

A simple pictorial diagram is shown in Fig. 6-3. Here, the circuit board is indicated as a rectangle and five components are placed on it at various points. The circuit layout is a general one, showing the transformer at the left-hand side of the board and the components arranged slightly right of center. The transformer is shown near the edge of the board because its output leads will feed a speaker. If the transformer was in the center of the board, an extra length of wiring would be required to access the speaker. The purpose of most pictorial drawings of electronic circuits is to indicate how the components can be placed to limit the interconnecting wiring. Extra wire means extra circuit resistance, as well as increased inductance and capacitance.

The schematic diagram for the pictorial drawing discussed here indicates that C1 and C2 are connected together

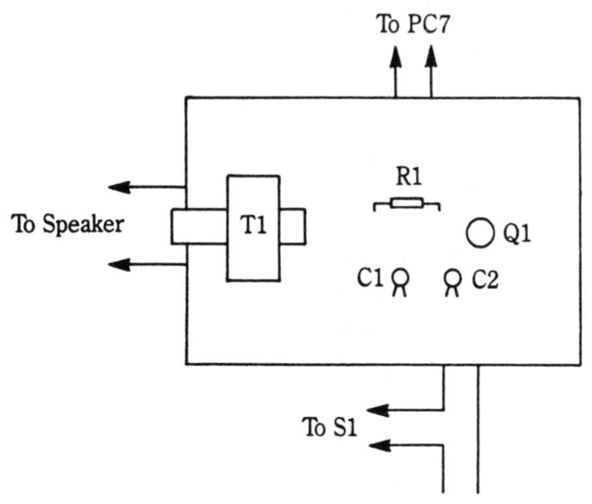

6-3 Pictorial diagram of a simple electronic circuit.

at one point. C1 and C2 are shown in close physical proximity as a result of this. If C1 was moved to the far left-hand side of the board, then an extra length of wire would have to connect the two. The extra resistance could affect the circuit's operation. Pictorial diagrams, then, are usually based on good electronic building practice, which stipulates that component lead wiring must be kept to an absolute minimum.

The examples of pictorial diagrams shown can be classified as *pictorial block drawings*. They do show the physical relationships and more closely symbolize the electronic components, but they do not show a true-to-life picture of the finished circuit.

When it becomes necessary to use a pictorial drawing to locate specific electronic components, a presentation similar to the one shown in Fig. 6-4 can be used. This highly complex piece of equipment might use hundreds of different components on each circuit board. Trying to locate one of these components on a circuit board can be very difficult, so the board is drawn actual size and an identification grid is placed around it. In this example, the circuit board is sectioned off from 1 to 8 vertically and in two sections (A and B) horizontally. A chart, which shows the component's designation and its grid location, is included. Locating C1 (at the top of the chart) involves examining the area in the grid of B2. All you do is go to the right-hand section of the circuit board (the B grid) and then look in the area encompassed by the number 2 grid. C1 can readily be identified, although it might take a bit of looking. Some schematic diagrams will also be handled in this manner.

Notice that in lieu of drawing the components in this particular example, some are indicated only by their alphabetic/numeric designation. A line is drawn on both sides of this designation to indicate the length of the component and, in part, the space it occupies.

Sometimes even this is not enough. Complex circuits might require a highly accurate line drawing of every aspect of all components used for their construction. In this case, many manufacturers simply take a closeup, overhead photograph of

Ref Desig	Grid Loc	Ref Desig	Grid Loc	Ref Desig	Grid Loc	Ref Desig	Grid Loc	Ref Desig	Grid Loc
C1	B2	C45	A6	Q10		R34	A3	R78	B7
C2	B2	C46	A6	Q11		R35	A4	R79	B7
C3	B2	C47	A7	Q12	B5	R36	A4	R80	B7
C4	B2	C48	B7	Q13		R37	A4	R81	B8
C5	B3	C49	A7	Q14	A5	R38	B4	R82	B7
C6	B3	C50	A7	Q15	A5	R39	A4	R83	B7
C7	B3	C51	A7	Q16	B5	R40	A4	R84	B7
C8	B3	C52	A8	Q17		R41		R85	B7
C9	B3	C53		Q18		R42		R86	B7
C10	B3	C54		Q19	B8	R43	B3	R87	A1
C11	B4	C54	A4	Q20	A8	R44	A4	R88	A6
C12	B4	C56	A4	R1	B2	R45	A4	R89	B6
C13	B3	C57	B6	R2	B2	R46	A4	R90	B6
C14	B4	C58	B6	R3	B2	R47		R91	A6
C15	A2	CR1	B3	R4	B3	R48		R92	A6
C16	A2	CR2	B3	R5	B3	R49	B5	R93	A6
C17	A2	CR3	B4	R6	B4	R50		R94	A6
C18	A2	CR4		R7	B4	R51	B5	R95	A6
C19	A2	CR5	A2	R8	B4	R52	B5	R96	A6
C20	A2	CR6	A2	R9	B4	R53	A5	R97	A6
C21	A3	CR7	A2	R10	B4	R54	A5	R98	A6
C22	A3	CR8	A2	R11	B4	R55	A5	R99	A6
C23	A3	CR9		R12		R56	A5	R100	A7
C24	B3	CR10	A2	R13		R57	B5	R101	A7
C25	A3	CR11	A4	R14	A2	R58	B5	R102	A6
C26	A3	CR12	A4	R15	A2	59	B5	R103	A7
C27	A4	CR13	B8	R16	A2	R60	B5	R104	A7
C28	B6	CR14	A6	R17	A2	R61	B5	R105	A7
C29	B5	CR15	A6	R18	A2	R62		R106	A7
C30		CR16	A7	R19	A3	R63	B6	R107	A7
C31	B5	L1	B2	R20	A2	R64	B6	R108	A7
C32	B5	L2	B2	R21	A2	R65	B6	R109	A7
C33	A5	L3	B3	R22	A2	R66	B6	R110	A7
C34	A5	L4	B3	R23	A4	R67	B6	U1	A1
C35	B6	L5	A3	R24	A3	R68	B6	U2	A5
C36	B7	Q1	B3	R25	A3	R69	B6	U3	B6
C37	B7	Q2	B4	R26	A3	R70	B6	U4	B7
C38	B7	Q3	Q3	R27	B2	R71	B6	U5	A6
C39	B7	Q4		R28	A2	R72	B7	U6	A7
C40	B7	Q5	B2	R29		R73	B7		
C41	B7	Q6	A2	R30		R74	B7		
C42	B7	Q7	A3	R31	B3	R75	B7		
C43	B7	Q8	A3	R32	A3	R76	B7		
C44	B7	Q9	A3	R33	A3	R77	B7		

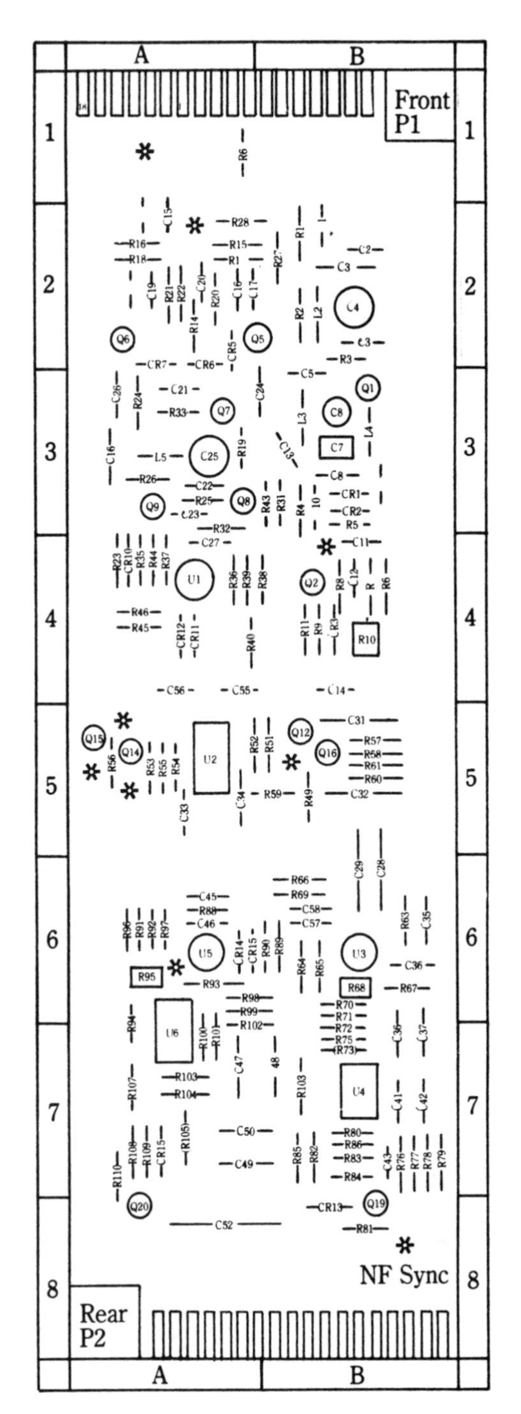

6-4 Pictorial diagram that includes a location chart.

the circuit board and then type in the various component designations. An alternate method involves making a line drawing like the one shown in Fig. 6-5. The detail here is excellent

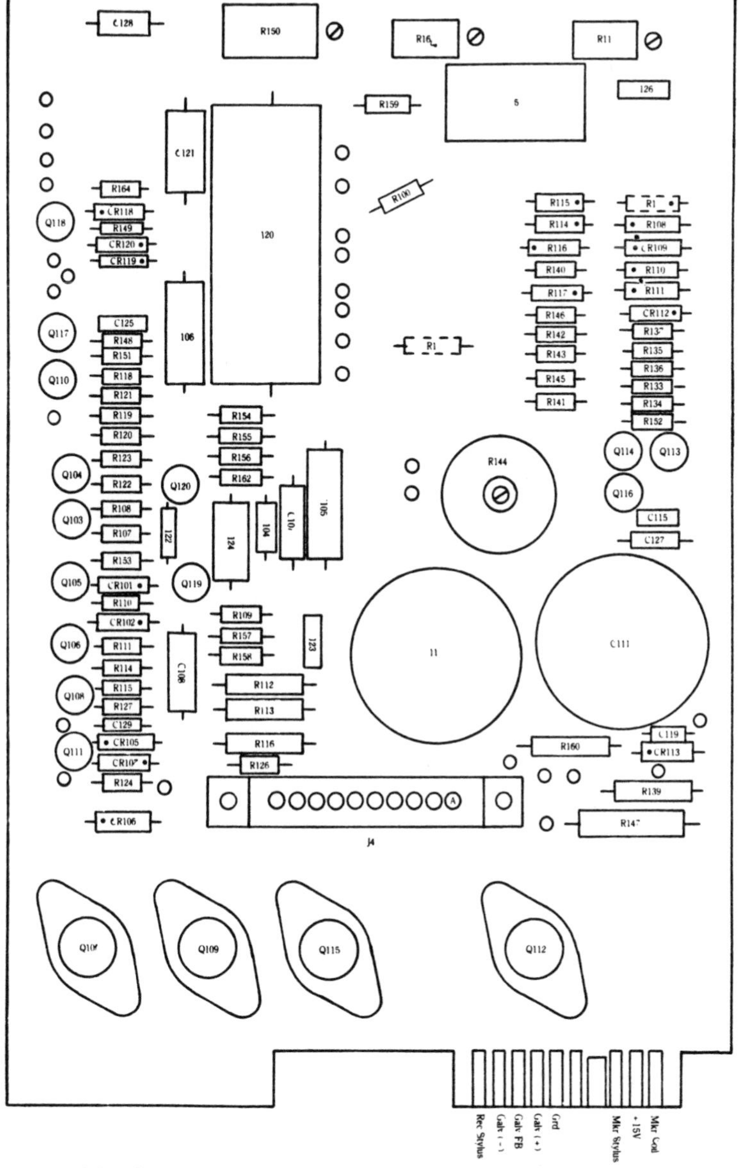

6-5 An accurate line drawing pictorial of a complex circuit.

and this drawing is a very close representation of the photograph. Some of these drawings are even prepared from photographs. Using a special developing process, the picture that was originally composed of halftones is processed down to black and white. In other words, the grays are eliminated. Making drawings of this type, either through a photographic process or by sheer artwork, is generally out of the realm of most hobbyists. However, it is quite easy to approximate circuit board components by drawing circles with a compass and straight lines with a ruler. The type of detail provided by extremely good artwork or photographic processes is almost never needed by the home experimenter.

The illustration in Fig. 6-6 shows the pictorial diagram of the first circuit introduced at the beginning of this book—the simple ac-to-dc converter whose block diagram is in Fig. 2-1 and the schematic is in Fig. 4-21. Since the schematic only consists of a few components, the pictorial diagram is small; it contains a resistor, two capacitors, a transformer, and the bridge rectifier, which in this case is contained within an integrated circuit.

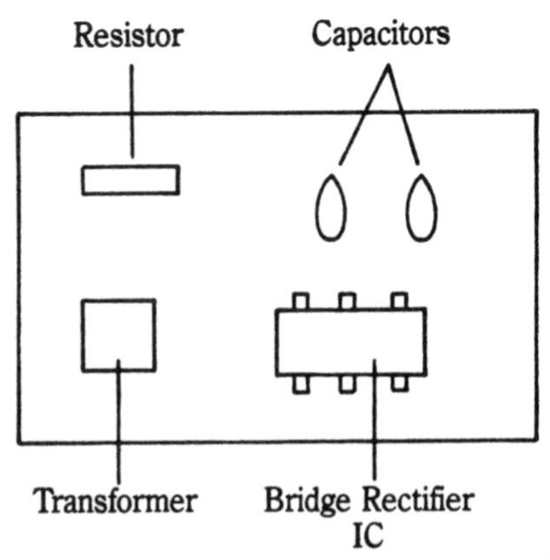

6-6 Pictorial representation of the schematic of Fig. 4-21.

To help you understand this illustration conversion process more fully, Fig. 6-7 shows how a simple schematic can be converted to a pictorial. Each connection in the simple resistor circuit shown at left (the schematic) must match up with

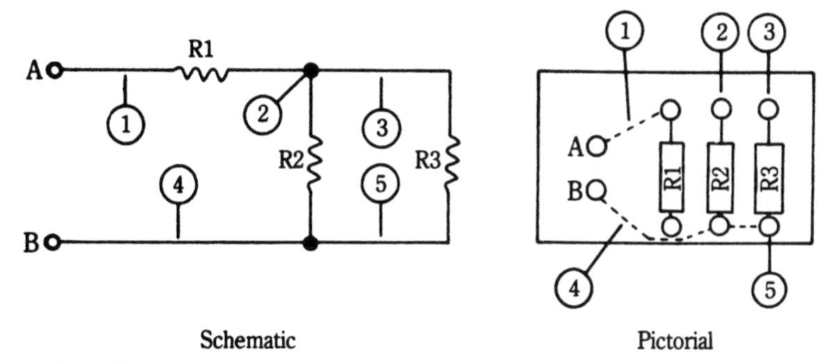

6-7 Example of how a circuit goes from schematic to pictorial form.

the pictorial layout shown at right. The pictorial is designed in such a way as to aid in circuit-board construction. Note how the circled numbers in the schematic correspond with the same points in the pictorial.

Incidentally, notice the component placement in the pictorial. The resistors are placed as close as possible together to reduce the bulkiness of the circuit; also, they are parallel to promote compactness. However, some components that generate a lot of heat, such as power transistors and large transformers, require more room around them to allow for proper cooling. The paths of a printed circuit must not be permitted to cross one another, and this layout facilitates this requirement. As you can probably imagine, circuit board design can get quite complicated and tedious with large, complicated circuitry.

Recall the schematic of the strobe light circuit in Fig. 5-16. The diagram in Fig. 6-8 shows the pictorial. The same basic techniques are followed to design the most compact, efficient, and functional layout as possible to eliminate extra costs and complications.

Three-dimensional drawings

Although not commonly drawn by electronic experimenters and hobbyists, three-dimensional simulations are sometimes used by commercial outfits to display their electronic circuits. They come in handy when it is necessary to represent mechanical functions and whole pieces of equipment. Most

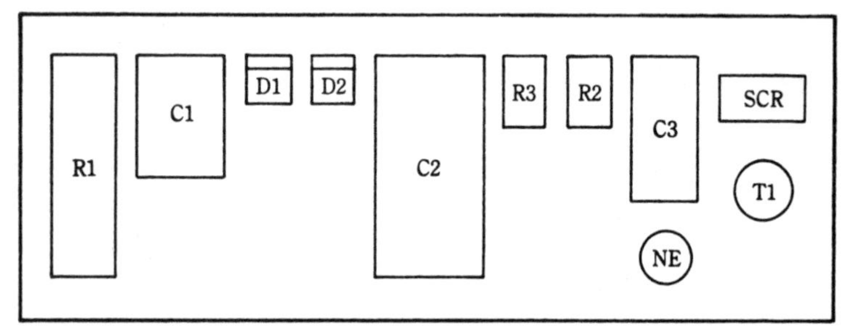

6-8 Pictorial representation of the schematic of the strobe light circuit in Fig. 5-16.

electronic circuits, however, can be adequately represented using two-dimensional drawings. In some cases where complex equipment is to be built, three-dimensional drawings can aid the builder in properly positioning components, especially where layered construction is involved.

This book has taken three-dimensional drawings into account to allow you to know how and why they are used. However, we are getting into a totally different realm that does not lend itself readily to the beginner. A person who has had no experience whatsoever with schematic diagrams can quickly learn to read and draw them. Pictorial diagrams can be read even more quickly, but drawing them involves the development of artistic ability that is far beyond the scope of this book. If you intend to make electronic diagramming your profession, then you are urged to take some basic art courses at a local college. Courses in electronics drafting will also be of a great deal of assistance.

Summary

The complex field of electronics not only lends itself to, but also requires many different types of symbology diagramming. The pictorial diagram is a highly useful illustration when dealing with assembly and service work. It cannot, however, be used very efficiently without a matching schematic diagram to serve as a cross-reference. The servicing of electronic equipment often involves consulting the schematic diagram first, then pulling in the pictorial diagram in order to locate the physical portion of the circuit where the schematic portion can be found. Only when the two are combined do the electronic and physical relationships align themselves into a complex language that aids the overall understanding.

Appendix A
Schematic symbols

AMPLIFIERS

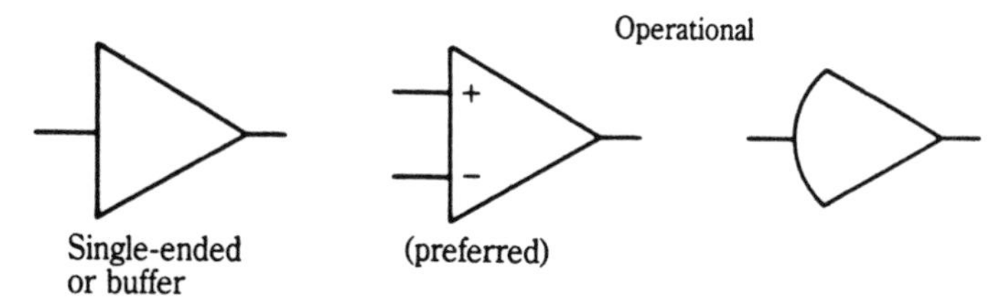

ANTENNAS

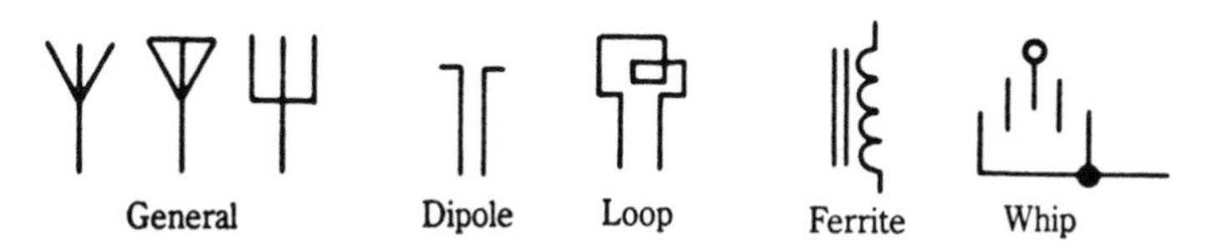

BATTERIES

Single cell

Multicell

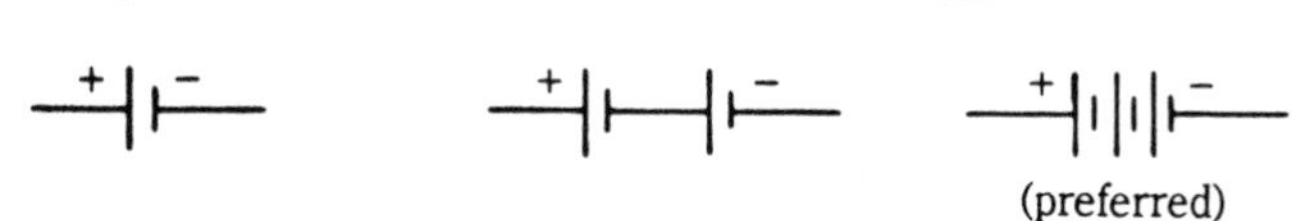

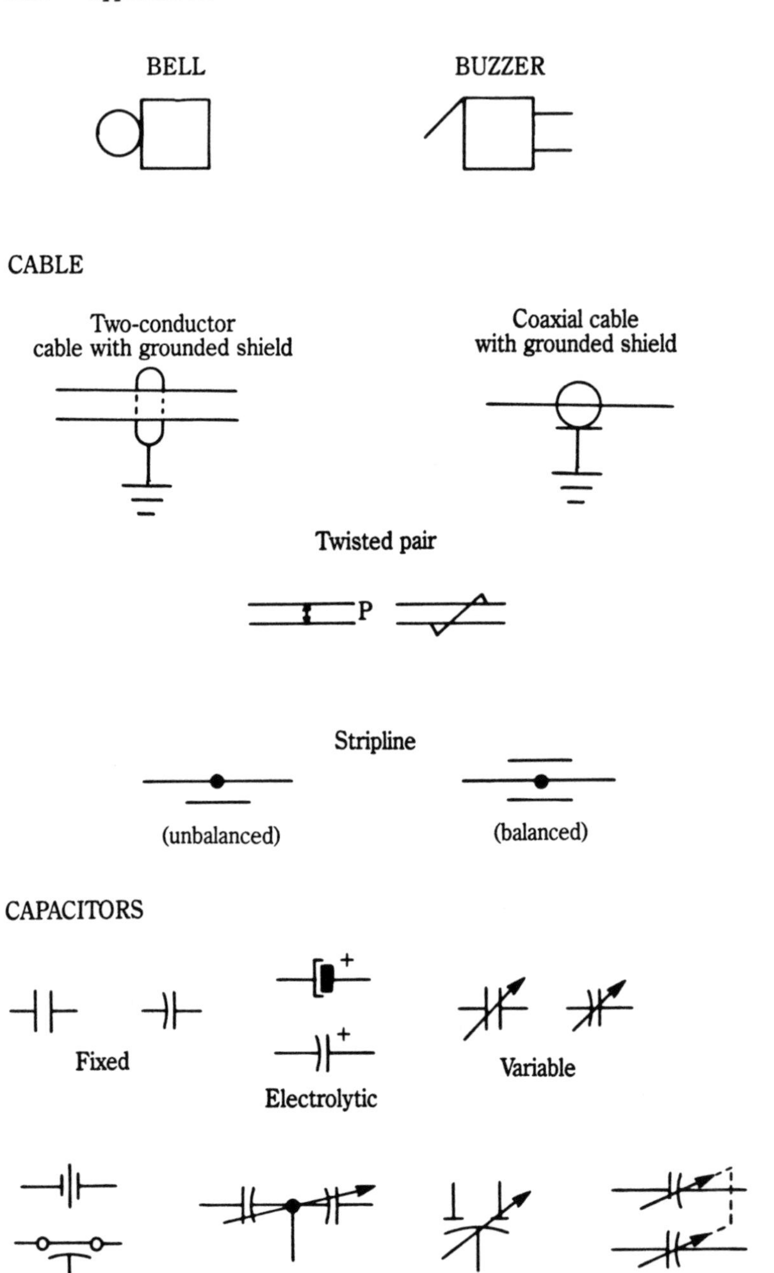
BELL
BUZZER
CABLE
Two-conductor cable with grounded shield
Coaxial cable with grounded shield
Twisted pair
P
Stripline
(unbalanced)
(balanced)
CAPACITORS
+
+
Fixed
Electrolytic
Variable
Feed through
Split stator
Split rotor
Ganged variable

CIRCUIT PROTECTORS

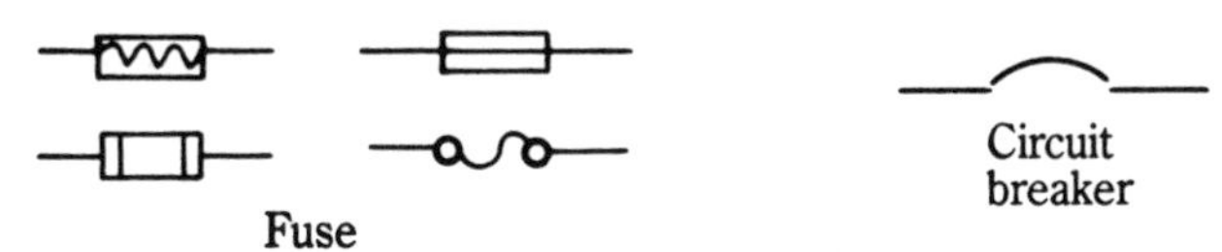

CONNECTORS

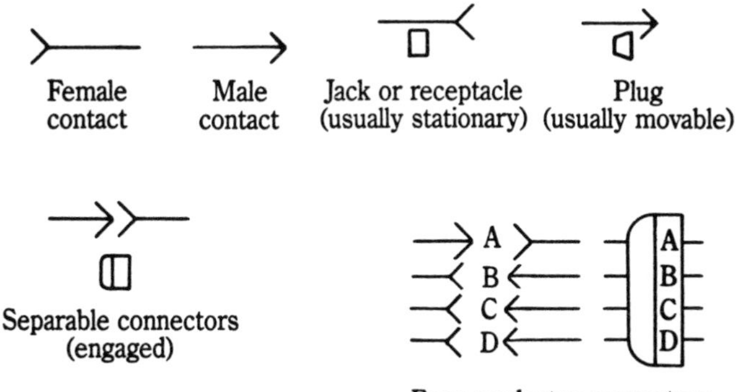

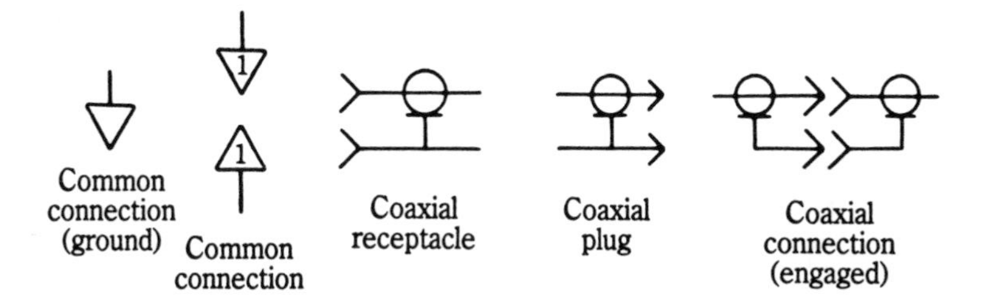

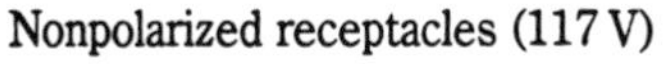

Nonpolarized receptacles (117 V)

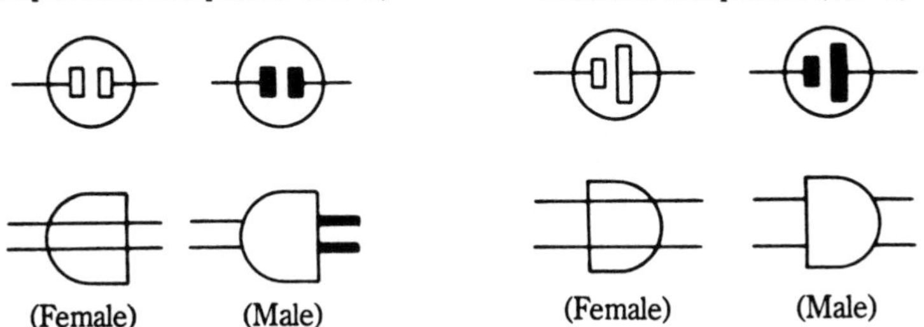

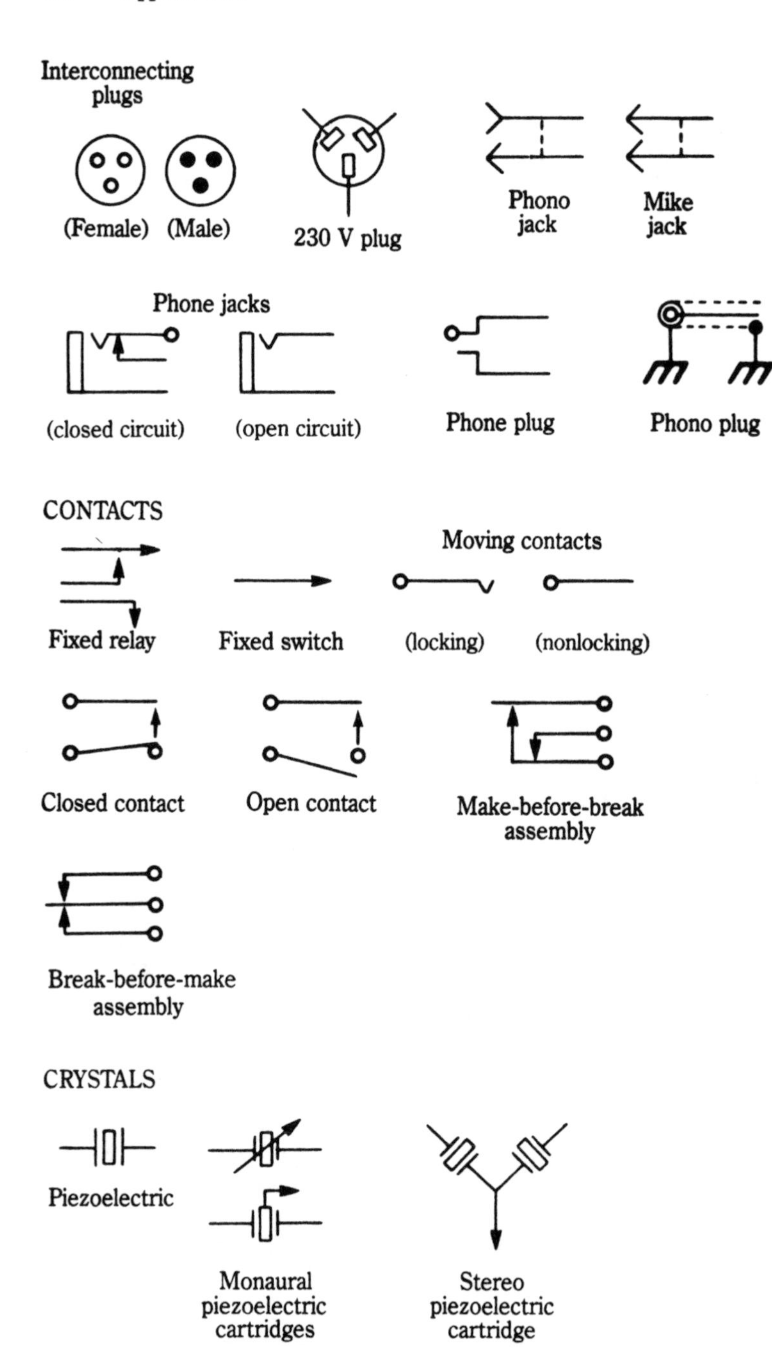
Interconnecting plugs
(Female)
(Male)
230 V plug
Phono jack
Mike jack
Phone jacks
(closed circuit)
(open circuit)
Phone plug
Phono plug
CONTACTS
Moving contacts
Fixed relay
Fixed switch
(locking)
(nonlocking)
Closed contact
Open contact
Make-before-break assembly
Break-before-make assembly
CRYSTALS
Piezoelectric
Monaural piezoelectric cartridges
Stereo piezoelectric cartridge

DIODES

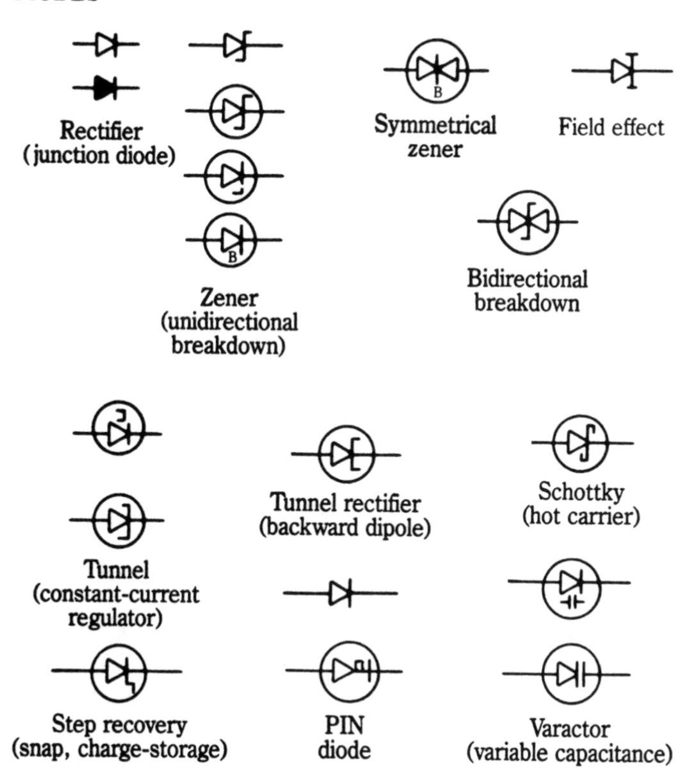

DIODES—OPTOELECTRONIC

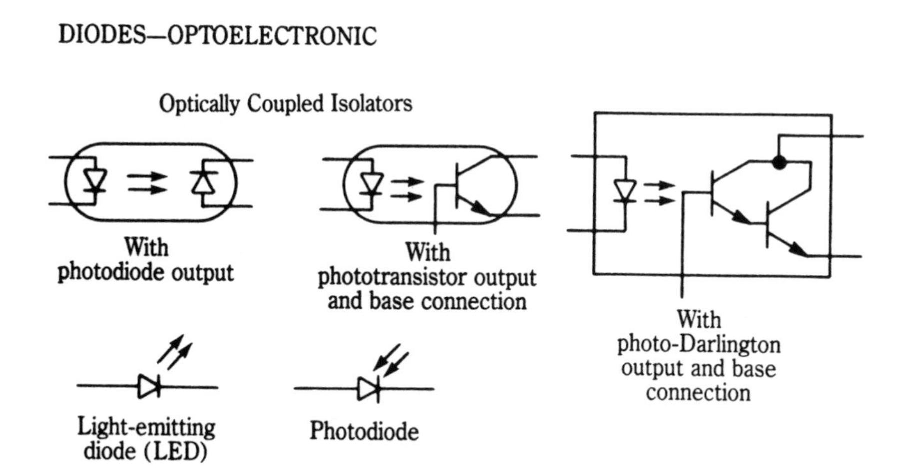

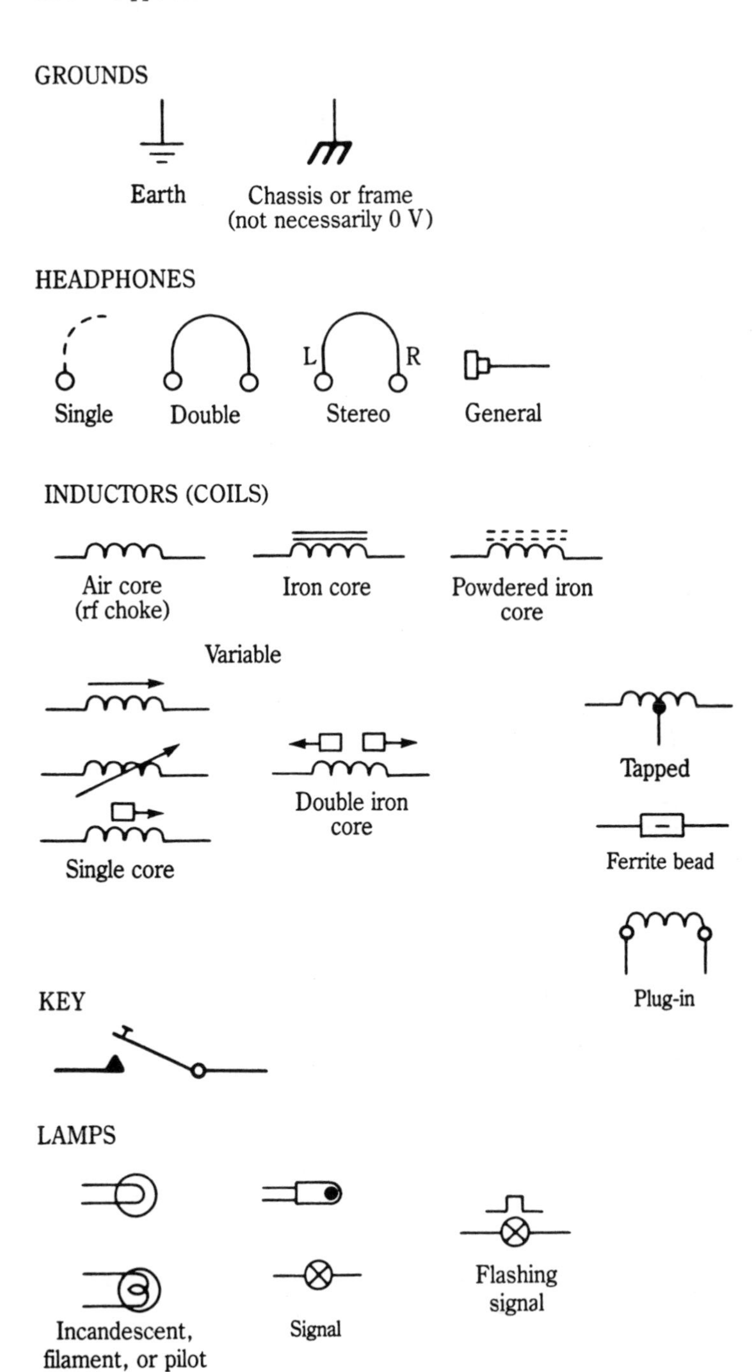
GROUNDS
Earth
Chassis or frame
(not necessarily 0 V)
HEADPHONES
Single
Double
L
R
Stereo
General
INDUCTORS (COILS)
Air core
(rf choke)
Iron core
Powdered iron
core
Variable
Single core
Double iron
core
Tapped
Ferrite bead
Plug-in
KEY
LAMPS
Incandescent,
filament, or pilot
Signal
Flashing
signal

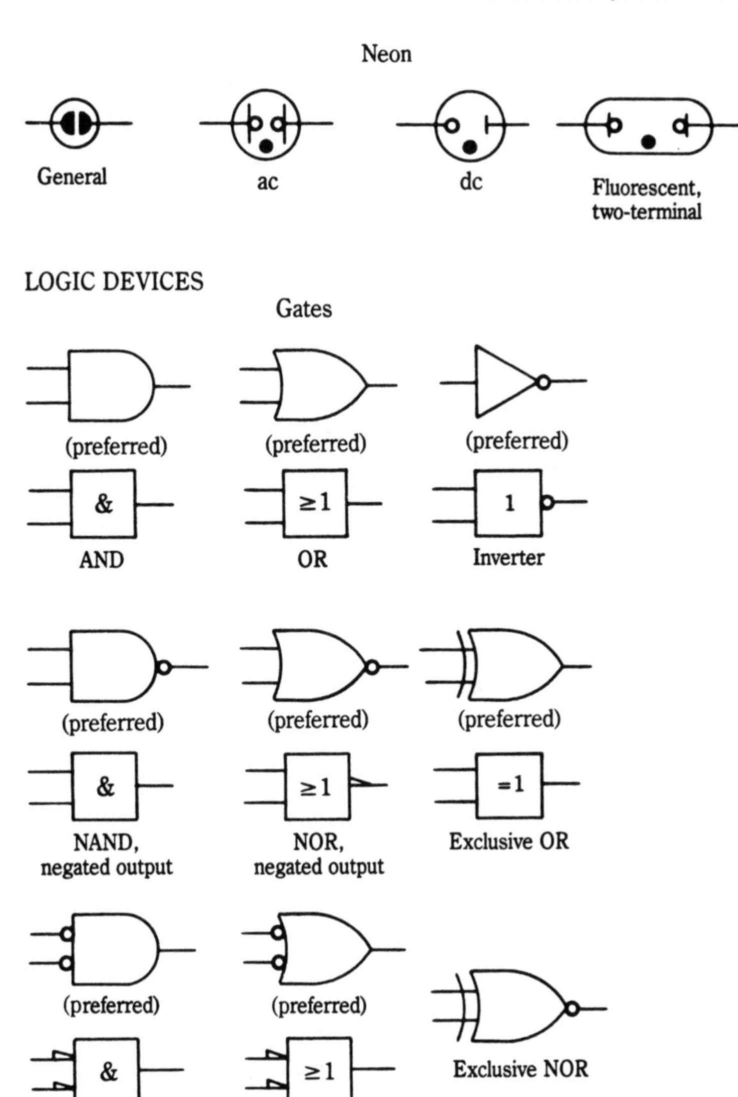
Neon
General
ac
dc
Fluorescent, two-terminal
LOGIC DEVICES
Gates
(preferred)
(preferred)
(preferred)
&
≥1
1
AND
OR
Inverter
(preferred)
(preferred)
(preferred)
&
≥1
=1
NAND, negated output
NOR, negated output
Exclusive OR
(preferred)
(preferred)
&
≥1
Exclusive NOR
NAND, negated inputs
NOR, negated inputs

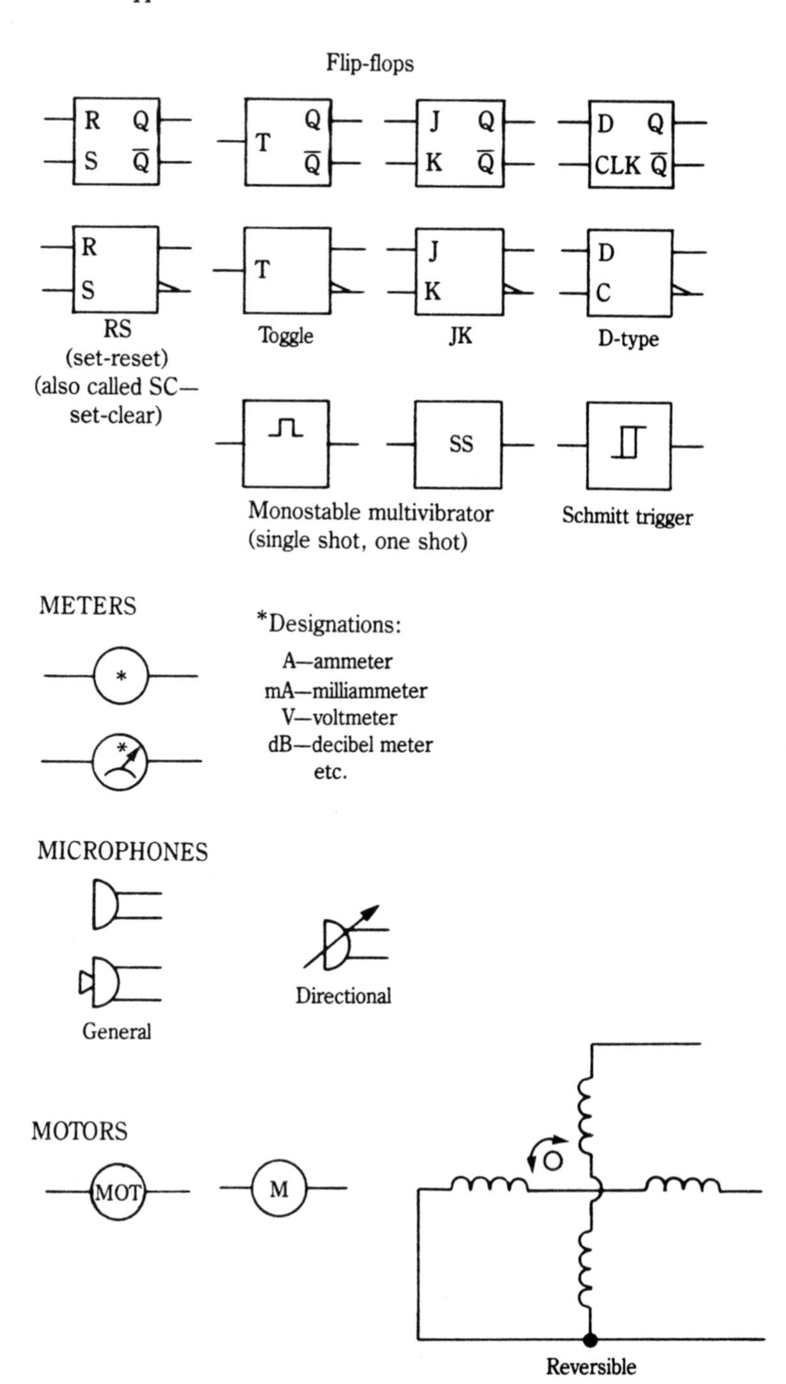
Flip-flops
R Q
S Q̄
T Q
Q̄
J Q
K Q̄
D Q
CLK Q̄
R
S
T
J
K
D
C
RS
(set-reset)
(also called SC—
set-clear)
Toggle
JK
D-type
SS
Monostable multivibrator
(single shot, one shot)
Schmitt trigger
METERS
*
*Designations:
A—ammeter
mA—milliammeter
V—voltmeter
dB—decibel meter
etc.
MICROPHONES
General
Directional
MOTORS
MOT
M
Reversible

RESISTORS

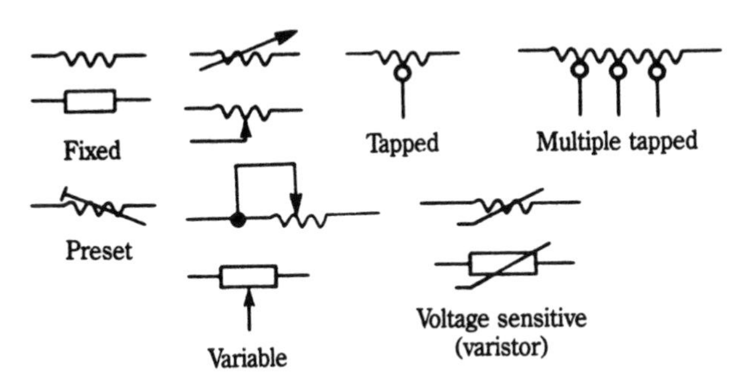

SPEAKERS

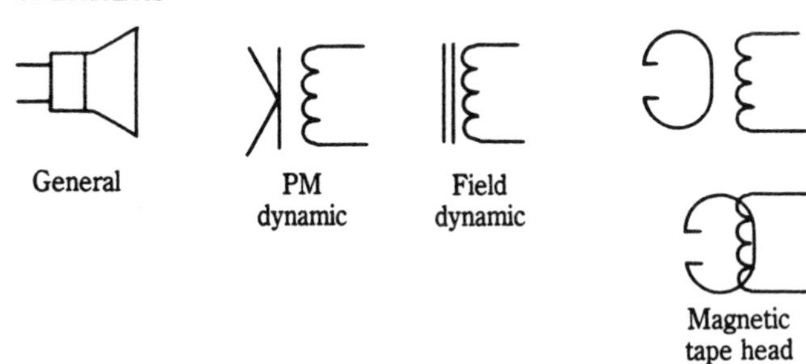

SOURCES

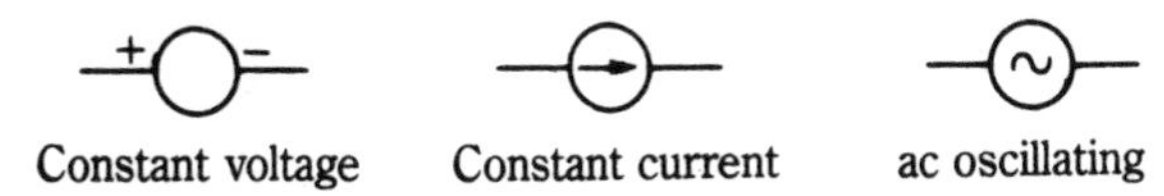

SWITCHES AND RELAYS

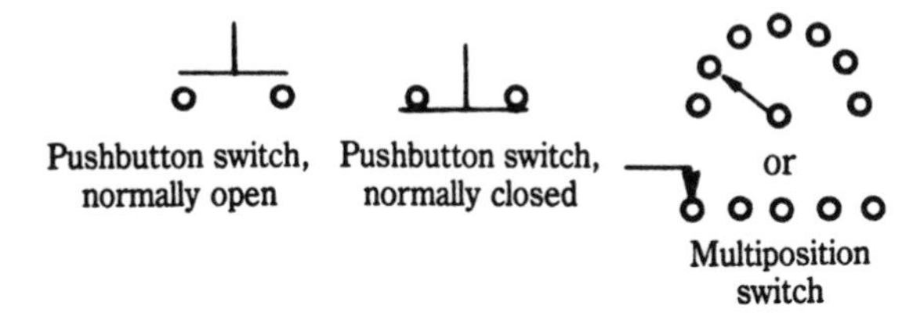

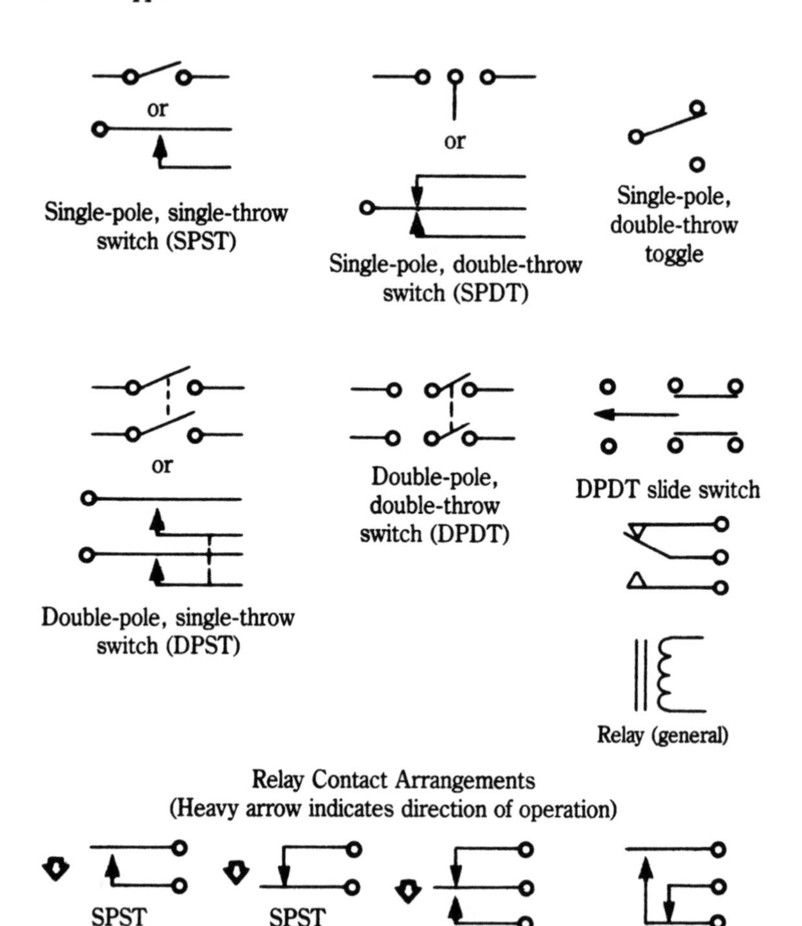

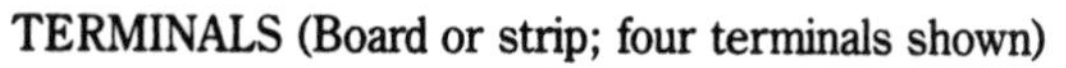

TEST BLOCK (Terminals shown)

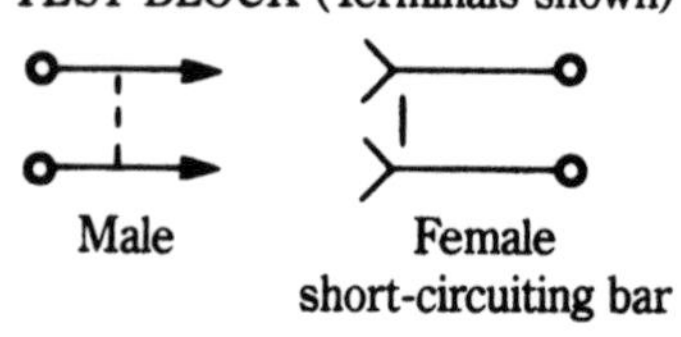

THYRISTORS

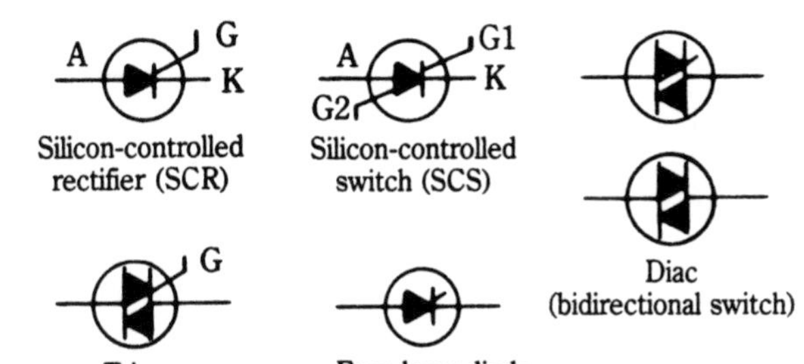

Triac (gated bidirectional switch)

Four-layer diode (PNPN or Shockley)

TRANSFORMERS

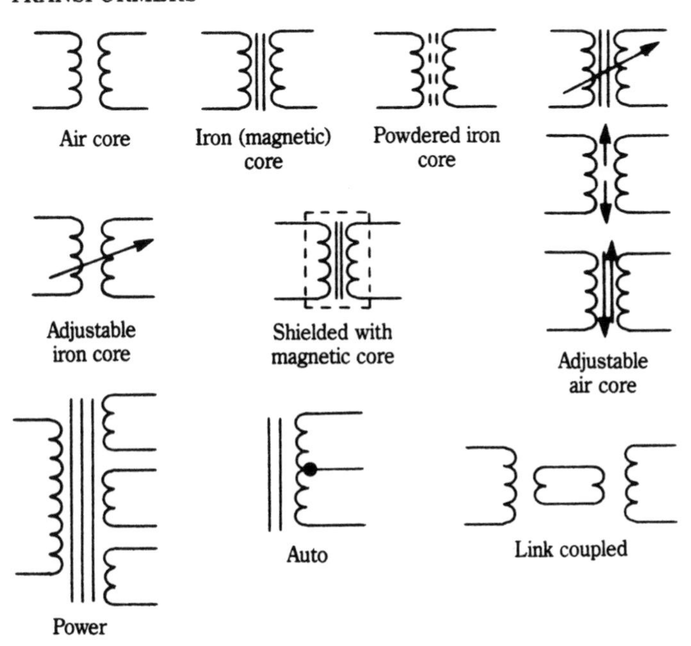

TRANSISTORS

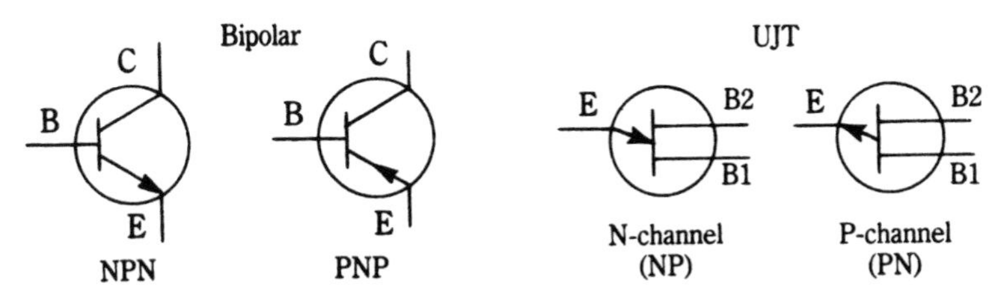

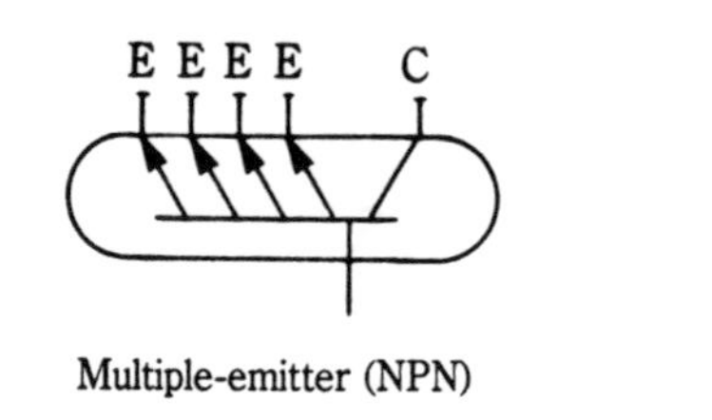

Multiple-emitter (NPN)

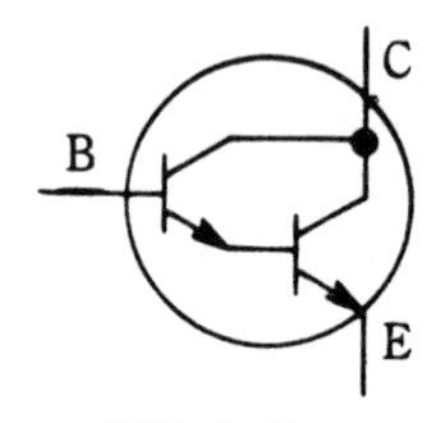

NPN Darlington

Hook (conjugate-emitter connection)

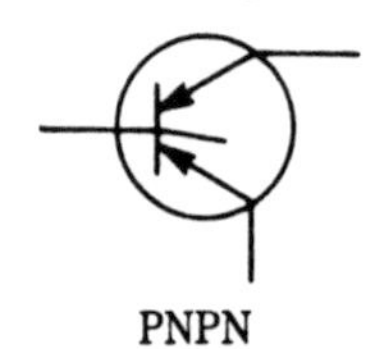

PNPN

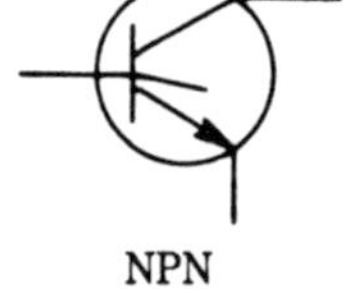

NPN

Tetrode

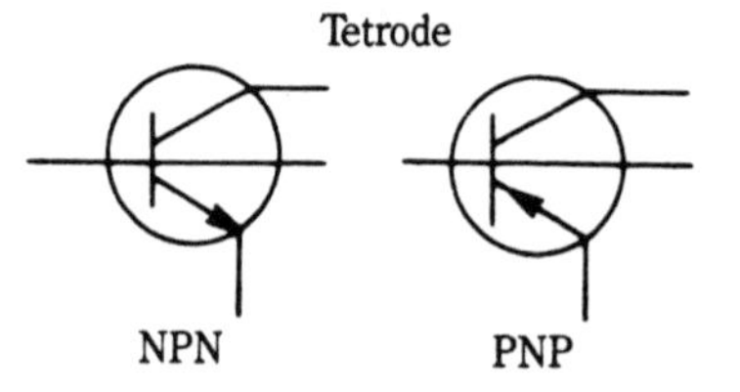

NPN

PNP

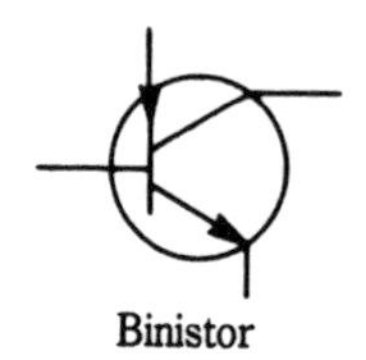

Binistor

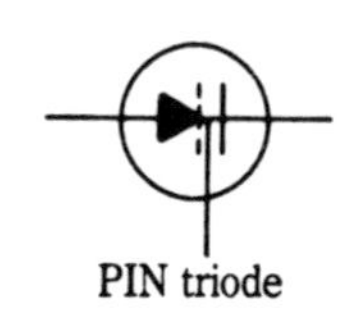

PIN triode

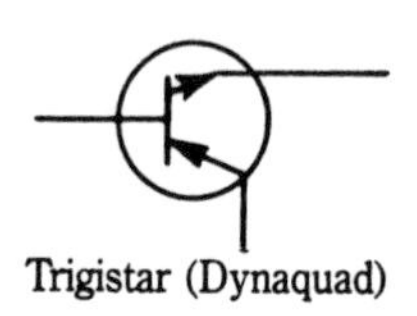

Trigistar (Dynaquad)

NPN Phototransistor

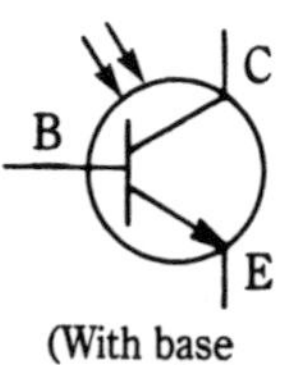

(With base connection)

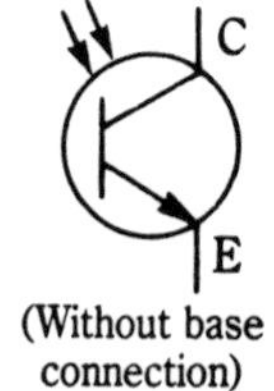

(Without base connection)

TRANSISTORS—FIELD EFFECT, N-CHANNEL
(for P-CHANNEL, reverse arrow direction)

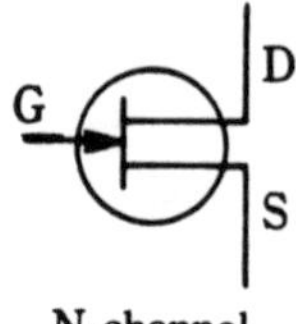

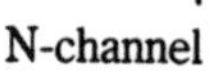

N-channel

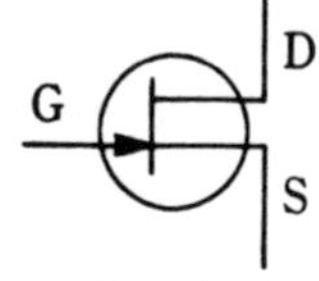

Junction (JFET)

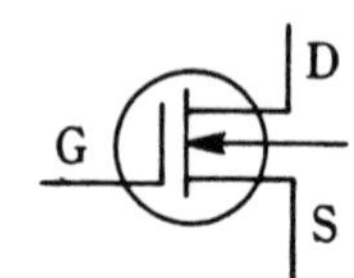

Three-terminal depletion (IGFET)

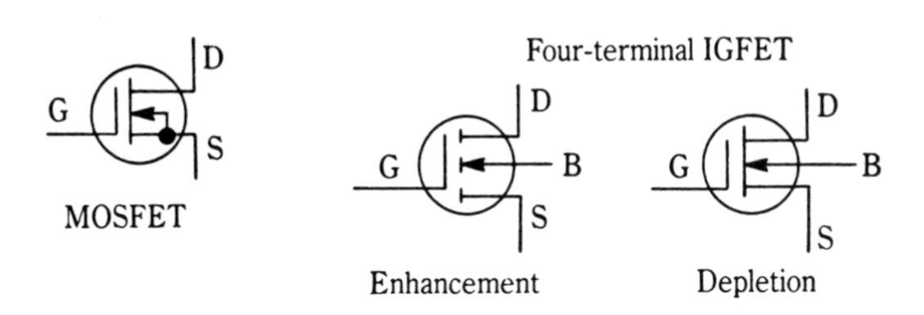

Five-terminal dual-gate IGFET

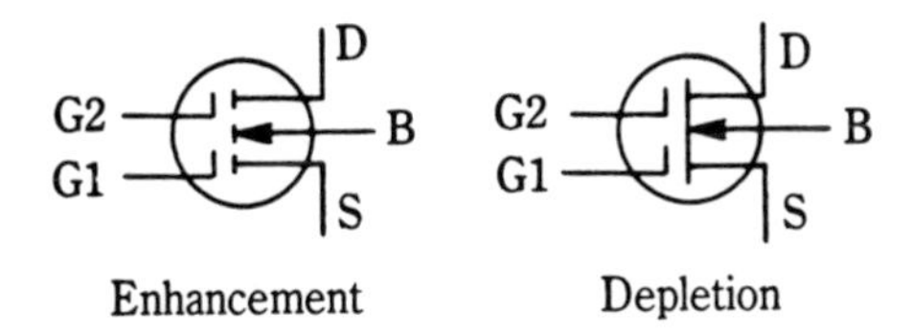

TUBE ELEMENTS

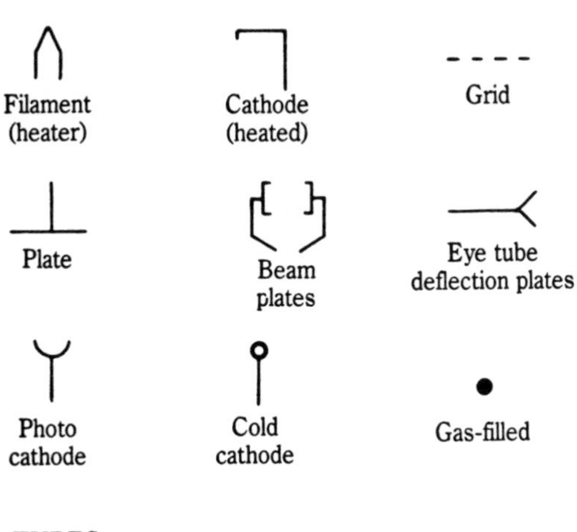

TUBES

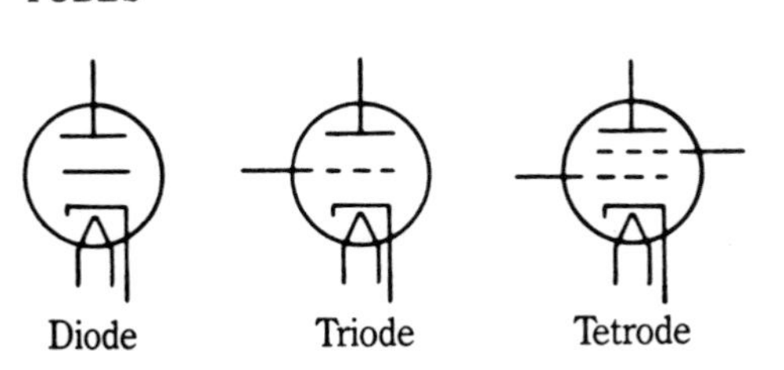

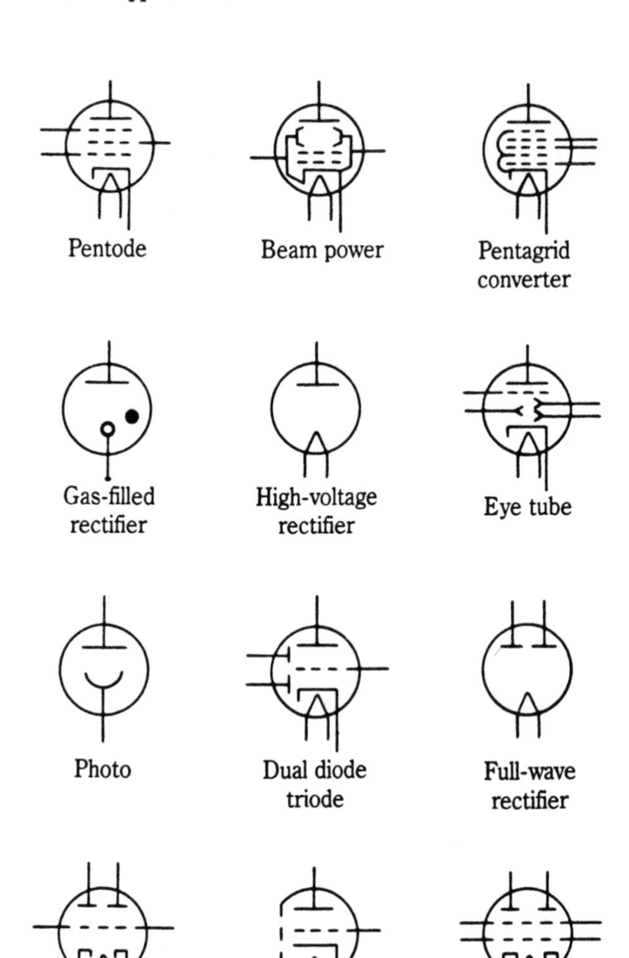
Pentode
Beam power
Pentagrid converter
Gas-filled rectifier
High-voltage rectifier
Eye tube
Photo
Dual diode triode
Full-wave rectifier
Dual triode
Two-section triode
Dual tetrode

Cathode Ray Tubes

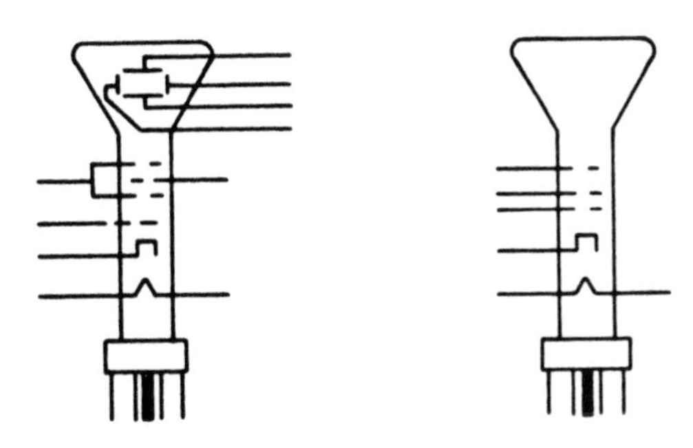

Electrostatic deflection

WAVEGUIDES

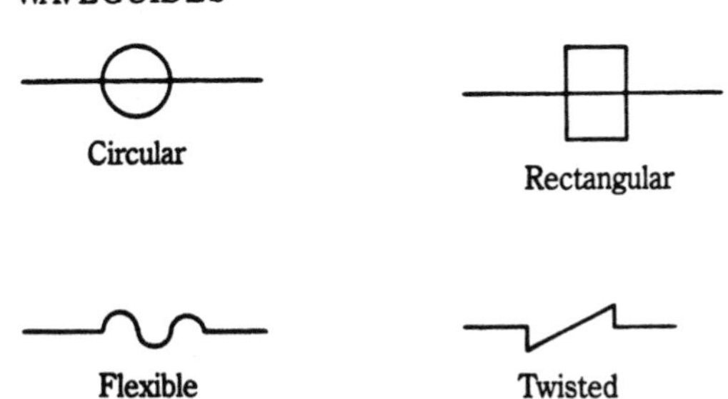

WIRING AND SHIELDING

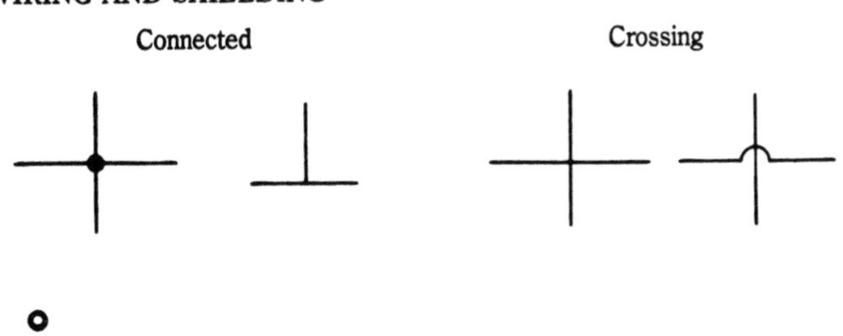

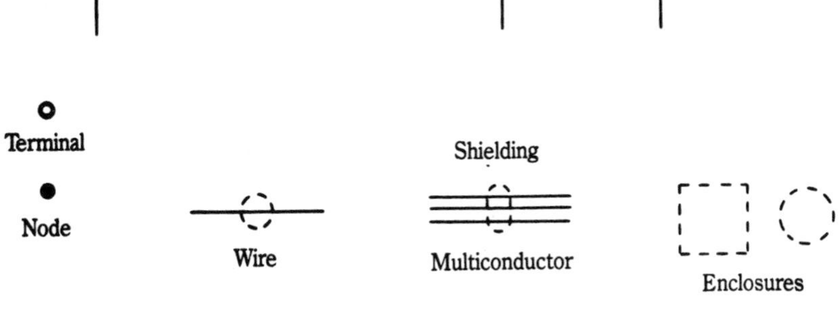

Appendix B
Resistor color codes

Color codes are standardized by the Electronic Industries Association.

Colors	Color Band Value (1st and 2nd bands)	Multiplier Band (3rd band)	Tolerance Band (4th band)
Black	0	1	
Brown	1	10	
Red	2	100	
Orange	3	1,000	
Yellow	4	10,000	
Green	5	100,000	
Blue	6	1,000,000	
Violet	7	10,000,000	
Gray	8	100,000,000	
White	9	1,000,000,000	
Gold	-	0.1	5%
Silver	-	0.01	10%
No band	-	-	20%

Examples of Resistor Color Codes

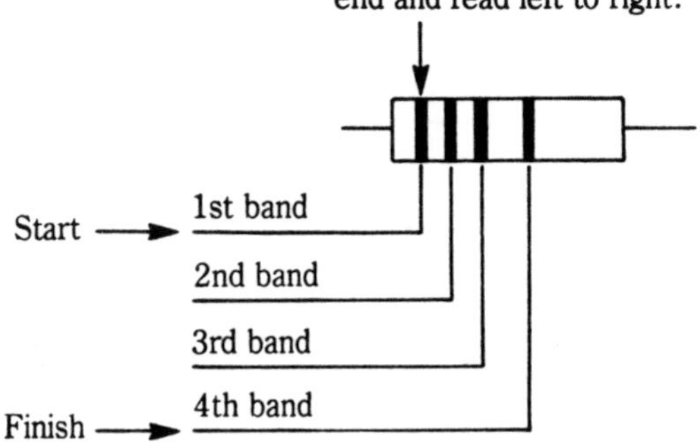

Example 1

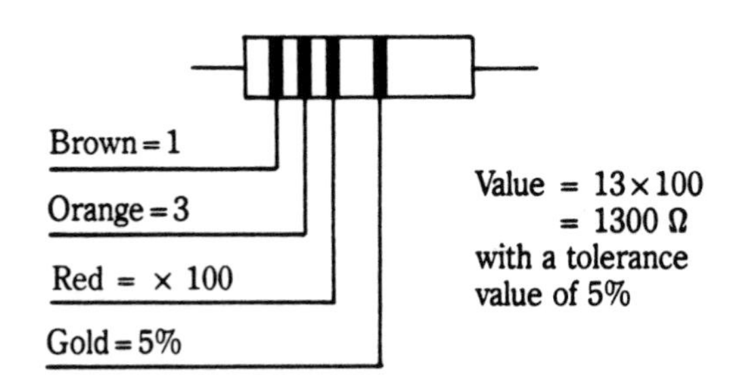

Example 2

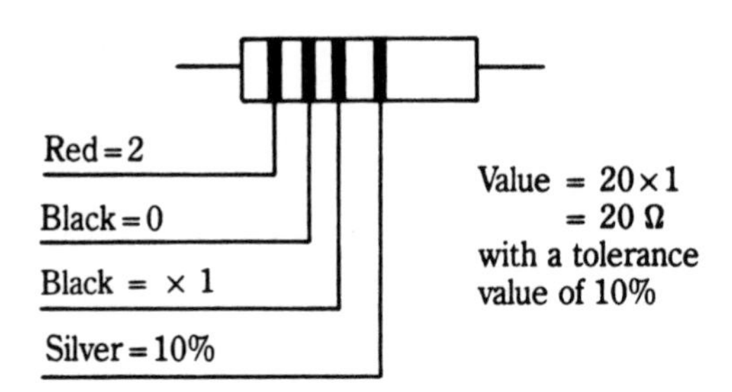

Index

Other Bestsellers of Related Interest

Encyclopedia of Electronic Circuits, Volume 1
—Rudolf F. Graf
Reference tool for the hobbyist, technician, student and design professional that contains 1,300 circuit designs, divided into 100 categories and meticulously indexed.
ISBN 0-8306-1938-0, #157328-3 $34.95 Paper
ISBN 0-8306-0938-5, #157322-4 $60.00 Hard

The Illustrated Dictionary of Electronics
6th Edition
—Stan Gibilisco
The most comprehensive up-to-date survey of electronics and computer technology. The #1 selling reference of its kind, with definitions for over 27,500 terms, abbreviations, and acronyms—more than any other electronics dictionary.
ISBN 0-8306-4396-6, #023599-6 $32.95 Paper
ISBN 0-8306-4397-4, #023598-8 $39.95 Hard

Encyclopedia of Electronics, 2nd Edition
—Gibilisco-Sclater
The first edition has sold 24,000 copies, with excellent reviews. The new edition has 40% new material, with new illustrations throughout. Painstaking review by experts from MIT and elsewhere assure its utility and accuracy. An essential reference useful to anyone connected to or interested in electronics.
ISBN 0-8306-3389-8, #156319-9 $69.50 Hard

Understanding Electronics, 3rd Edition
—R.H. Warring and G. Randy Slone
#1 bestselling electronics title revised and updated. Explains electronics and how electronic components work without extensive use of formulas, yet gives in-depth explanations and examples.
ISBN 0-8306-9344-0, #157376-3 $13.95 Paper

Build Your Own Laser, Phaser, Ion Ray Gun
—Iannini
For electronics hobbyists and sci-fi enthusiasts, this perennial bestseller includes 30 projects, complete with schematics and assembly instructions, for building high-power lasers, ultrasonic devices, magnetic distortion detectors, and more.
ISBN 0-8306-0604-1 #156069-9 $18.95 Paper